Computational Social Sciences

Computational Social Sciences

A series of authored and edited monographs that utilize quantitative and computational methods to model, analyze and interpret large-scale social phenomena. Titles within the series contain methods and practices that test and develop theories of complex social processes through bottom-up modeling of social interactions. Of particular interest is the study of the co-evolution of modern communication technology and social behavior and norms, in connection with emerging issues such as trust, risk, security and privacy in novel socio-technical environments.

Computational Social Sciences is explicitly transdisciplinary: quantitative methods from fields such as dynamical systems, artificial intelligence, network theory, agentbased modeling, and statistical mechanics are invoked and combined with state-ofthe-art mining and analysis of large data sets to help us understand social agents, their interactions on and offline, and the effect of these interactions at the macro level. Topics include, but are not limited to social networks and media, dynamics of opinions, cultures and conflicts, socio-technical co-evolution and social psychology. Computational Social Sciences will also publish monographs and selected edited contributions from specialized conferences and workshops specifically aimed at communicating new findings to a large transdisciplinary audience. A fundamental goal of the series is to provide a single forum within which commonalities and differences in the workings of this field may be discerned, hence leading to deeper insight and understanding.

More information about this series at http://www.springer.com/series/11784

Jeff Collmann · Sorin Adam Matei
Editors

Ethical Reasoning in Big Data

An Exploratory Analysis

Editors
Jeff Collmann
Georgetown University
Washington, DC
USA

Sorin Adam Matei
Purdue University
West Lafayette, IN
USA

Computational Social Sciences
ISBN 978-3-319-28420-0 ISBN 978-3-319-28422-4 (eBook)
DOI 10.1007/978-3-319-28422-4

Library of Congress Control Number: 2016933232

Printed on acid-free paper

This Springer imprint is published by Springer Nature
The registered company is Springer International Publishing AG Switzerland

To Lilliam and Daniela

Acknowledgments

We owe the flow of ideas, words, and digital bits in this project to many people. First, we wish to acknowledge and thank the National Science Foundation for sponsoring our workshop, "Privacy in the Infosphere: an NSF-sponsored workshop on ethical analysis of big data," award number 1502325 which enabled us to assemble a rich, multidisciplinary, multi-institutional, and multi-sectoral group at Georgetown University for articulation and vigorous discussion of many of the ideas expressed in these chapters. From Georgetown, we first thank Spiros Dimolitsas, Senior Vice President for Research and Chief Technology Office, who sponsors the "big data" project that prompted our initial reflection. We also want to thank Robert Groves, Provost, Georgetown University and Ed Montgomery, Dean, McCourt School of Public Policy, for helping sponsor and welcome participants to the workshop.

From Purdue, we would like to thank Dr. Elisa Bertino, the head of the CyberCenter, whose leadership at Purdue created many interdisciplinary bridges and the Discovery Park Leadership, especially Dr. Alan Rebar who made possible the Discovery Park Fellowship Program, which supported part of Sorin Adam Matei's research activities that lead to his contributions to the book.

We also want to thank with special enthusiasm the scholars who came to our workshop including Michael Steinmann, Stevens Institute of Technology; Douglas Richardson, Association of American Geographers; Kevin T. FitzGerald, SJ, Georgetown University; James Giordano, Georgetown University; Andrew Russell, Stevens Institute of Technology; Rochelle Tractenberg, Georgetown University; John E. Marcotte, University of Michigan; Jon D. Miller, University of Michigan; Lisa Singh, Georgetown University; Isabel Bradburn, Virginia Tech; Howard Everson, CUNY; Ben Berkman, National Institutes of Health; Joel Kupersmith, Georgetown University; Lee Rainie, Pew Internet project; Jay Stanley, ACLU; Justin Knapp; Sylvia Mann Department of Health Hawaii; and Sam Garner, National Institutes of Health.

Book projects require special effort from family and friends who must live with disgruntled, distracted, and distressed loved ones. Special thanks to Lilliam Oliva Collmann and Daniela Matei.

Under award #SMA 1338507, NSF also sponsored work on forced migration that appears in several chapters. NSF Award BCS 1244708 on credibility and trust in social media also contributed to the conceptual work in Chap. "A Theoretical Framework for Ethical Reflection in Big Data Research." Any opinions, findings, and conclusions or recommendations expressed in this material are those of the authors and do not necessarily reflect the views of the National Science Foundation.

Contents

Introduction

Jeff Collmann and Sorin Adam Matei

The advent of the Internet and of mass digitization of research information processes brought about among many other things the ability to harvest, sometimes implicitly, a wealth of human behavioral, biological, economic, political, or social data. The emergence of social media further amplified this trend, as each post, like, share or comment can be turned into analyzable data. The sequencing of the human genome and the inventory of many basic molecular processes in the human body have further expanded the universe of information. It is estimated that 90 % of all existing data was generated in the last few years (Wall 2014). Furthermore, data typically arrives as a deluge, not as a trickle. In the previous decades, social research was limited to samples of hundreds of thousands of cases. Now, datasets include millions of records. Seen from this lens, data has acquired the attribute "big." This is, however, not only a quantitative attribute, but a qualitative one (Macy 2015). Big data refers often to populations and takes the form of complete counts. It is, at the same time, captured not only as attributes, but as relationships (Harris 2013). Sieving through the census-like inventories with high speed and highly efficient computing algorithms has made possible new discoveries in genetic and clinical medical research or in social scientific understanding of diffusion processes (Christakis and Fowler 2009).

Data collection on such a massive scale involving millions of individuals and data points was many times done through automated, technological means that left some human ethical concerns aside to be discussed after the fact. Such a situation is fraught with dangers, including major harm to the individuals observed. Their right to privacy, free expression, autonomy and their trust in the scientific establishment

J. Collmann (✉)
Georgetown University, Washington, DC, USA
e-mail: collmanj@georgetown.edu

S.A. Matei (✉)
Purdue University, West Lafayette, IN, USA
e-mail: smatei@purdue.edu

J. Collmann and S.A. Matei (eds.), *Ethical Reasoning in Big Data*,
Computational Social Sciences, DOI 10.1007/978-3-319-28422-4_1

or in government can be put in serious danger. Creating a framework for ethical reasoning that can be employed *before* the research process starts, thus, becomes an obvious priority for the research community.

This book grows from a multidisciplinary, multi-organizational and multi-sector conversation about the privacy and ethical implications of research in human affairs using big data. Authors include a wide range of investigators, practitioners and stakeholders in big data about human beings who also routinely reflect on the privacy and ethical issues of this phenomenon. Together with other colleagues, all participated in a workshop entitled "Privacy in the Infosphere: an NSF-sponsored workshop on ethical analysis of big data". The authors have several different perspectives on big data but all express caution in rushing to judgment about its implications. Their diversity suggests the stakes at hand in big data, especially the stakes for scientific research and science as a legitimate institution in our society. Yet, perhaps, because of its implications for individual privacy and the spectacular revelations about various uses of big data by government and commercial organizations, discussions about the ethics of big data have not remained solely the purview of specialists—the public takes an interest. Thus, the authors of this book write with their eyes and ears directed in multiple directions: toward their peers in scientific research, toward the government and philanthropic agencies who fund research, and toward the citizens of our society whose taxes underpin, and whose lives provide the information that constitutes the focus of much big data research.

We subtitled this book "an exploratory analysis" because we think that many ethical questions remain unanswered, indeed unasked, about big data research focused on human affairs. Our limited experience with big data in all its contemporary forms urges caution and humility but not inaction. We are all actively engaged in big data research about human affairs in one way or another and think that ethical reasoning about big data issues as they emerge from scientific practice is likely to produce more useful results than speculation in the absence of experience. Yet, controversy exists about the form, extent, context and formality of reasoning about big data ethical issues.

For example, participants in our privacy workshop did not all agree on several issues, such as the connection between data provenance and privacy concerns or the mechanisms by which privacy procedures should be designed. The workshop occurred on April 15–16, 2015 in the Philodemic Room of Georgetown University, Washington, DC. The participants included an organizing committee, a multidisciplinary set of researchers who have received NSF funding for big data projects from the Education and Human Resources Directorate (EHR) and the Building Capability and Community in Big Data (BCC) program, and a distinguished panel of big data stakeholders from a range of disciplines, organizations and community sectors, a composition intended to capture diverse perspectives and encourage lively discussion. The organizing committee prepared and sent materials to all participants before the workshop to help create a common foundation for launching the discussion. The materials included a white paper, a case study template and an example case study. After reviews of the concepts of big data and of privacy, the white paper focused on a heuristic device (the Privacy Matrix) designed to assist big data investigators in identifying privacy issues with potential ethical implications in their work (Steinmann, Shuster, Collmann, Matei, Tractenberg, FitzGerald, Morgan, and Richardson 2015 and Steinmann, Matei and Collmann below). Four participants prepared case studies based on their own experience.

The diversity of backgrounds, experiences and organizational affiliation manifested itself in a diversity of perspectives on the interpretation of the case studies, the implications of big data research for privacy and ethical analysis and ideas about Data Management Plans as tools for protecting privacy and promoting ethical reflection in big data research. Some participants argued that, in contrast to government or commercial uses of big data, big data research posed few new privacy or ethical issues and that existing mechanisms (including the existing guidance for NSF proposal Data Management Plans) sufficed to handle them. Adding new ethical concerns or human subject review questions might increase the bureaucratic overhead without enhancing the value of big data science. Other participants argued that big data research invites renewed reflection on privacy and ethical issues in scientific research. Although existing mechanisms might suffice in some cases, other cases such as those presented in the workshop challenge current practice and institutional procedures. In addition, some research communities such as computer scientists have begun work on big data about human subjects with little experience of the human subjects protection process or tradition. Until such time as the scientific community has developed more experience with big data research, individual scientists and their institutional research support groups should examine big data projects on a case-by-case basis using tools such as the Privacy Matrix (Steinmann, Shuster, Collmann, Matei, Tractenberg, FitzGerald, Morgan, and Richardson 2015 and Steinmann, Matei and Collmann below) or expanded Data Management Plan format (see Collmann, FitzGerald, Wu and Kupersmith below) to assist them in proactively anticipating issues while planning research and comprehensively analyzing issues as they arise in the course of research. This diversity of opinion echoes the current controversy over the Notice of Proposed Rulemaking for Revisions to the Common Rule (see http://www.hhs.gov/ohrp/humansubjects/regulations/nprmhome.html).

Dedicated to the practice of ethical reasoning and reflection in action (see DiEuliis and Giordano, Tractenberg below), the authors in this book offer a range of observations, lessons learned, reasoning tools and suggestions for institutional practice to promote responsible big data research on human affairs. The need to cultivate and enlist the public's trust in the ability of particular scientists and the institutions of science constitutes a major theme running throughout the book. When scandals develop about the misuse or breach in confidentiality of individually identifiable information in human subjects research, participants suffer various types of harm and science as an institution in our society sustains some loss of the public's trust and sense of legitimacy. Above all, as Rainie describes in his chapter on the results of Pew surveys on privacy in America, members of the public expect to grant permission for researchers to gather, use, and secondarily reuse data about them, especially individually identifiable data such as biospecimens.

1 Overview of the Chapters

The book examines ethical reasoning about big data from multiple perspectives and with reference to multiple specific domains. The opening chapter (Steinmann, Matei and Collmann) explains the approach to privacy that has informed our entire project, particularly the idea explained most cogently by Nissenbaum that the meaning of "privacy" depends on the context (Nissenbaum 2009; Steinmann et al. 2015). We added to her key insight the idea that the impact of privacy breaches and, thus, the significance of privacy protections depends on their effect with respect to the key ethical values of beneficence, non-malificence or do no harm, autonomy, social justice and trust. The intersection of specific privacy contexts with these five ethical values yields a privacy matrix, an initial tool offered to aid investigators in identifying the key privacy and ethical implications of their work and, thus, to adopt relevant, effective privacy protections. As we thought about the cases presented in the NSF conference, we realized that building big data for analysis can entail multiple processes and ethical concerns including what we labeled the 4Rs of reuse, repurposing, recombining and reanalysis. These concepts derive from the possibility that many big data sets come from sources other than the context of their analysis in a specific research project and might also later contribute data to projects other than the one of their origin. We refer to these twin possibilities as the ethical provenance of data collected in the past and the ethical horizon of data being collected in the present for potential use in the future. We pay special attention to the fact that data might originate in one context (e.g., government) and become incorporated into research in another context (e.g., Common Rule science). We observe that this context switching often has ethical implications—a point elaborated in chapter 10 that explains a decision tree for creating Data Management Plans.

The second section of the book examines the privacy and ethical implications of big data through the lenses of specific domains, including current attitudes about privacy among the American public (Rainie), using Code, Laws, Markets, and Norms for scientists and other Big Data analysts to build trust (Knapp), the ethical implications of current research in genomics (Berkman, Shapiro, Eckstein, and Pike) and neuroscience (DiEuliis and Giordano), and personal privacy on the Internet (Singh). From the perspective of privacy context, these chapters examine key institutional sectors (civil society, commerce, government, Common Rule science). A common ethical theme emerges across these chapters. Maintaining the trust of the American public underlies creating, analyzing and using big data in all these contexts; but, especially in scientific research where the traditions of avoiding harm and obtaining informed consent in the name of long-term human benefit underlies its fundamental legitimacy. This section contains individual chapters devoted to ethical reasoning in genomics and neuroscience, two disciplines that produce truly massive data sets. For example, Berkman et al. describe the rise of Large Scale Genomic Repositories (LSGRs) in genomics which, through the application of contemporary gene sequencing techniques and the recombination of multiple, large data sets, create resources with terabytes of data about more than one million research participants. Distinguishing between welfare and non-welfare harms, Berkman et al. highlight the importance of non-welfare harms by saying "maintaining trust in the research enterprise and in the process of developing LSGRs is fundamental to the ongoing success of LSGRs and the research enterprise. And yet, the way that LSGRs are currently being created falls short of best practices for establishing and fostering trust." (see Berkman, Shapiro, Eckstein, and Pike below).

After providing a detailed analysis of the critical role of big data in contemporary neuroscience, DiEuliis and Giordano emphasize the role of ethical discourse in enabling scientific progress in the field. They observe, "Such discourse must: (1) acknowledge and define the changing neuroscientific capabilities conferred by the use—and/or misuse—of big data approaches; (2) identify those neuroethico-legal and social issues generated by such use and effect(s); and (3) establish methods to address and resolve such issues, questions and problems, in part through both the development of (practice) guidelines, and by informing and contributing to public policy." (DiEuliis and Giordano). These two disciplines address questions about core elements of human identity, our genetic makeup and our brain with their twin implications for a range of behavioral, psychological and physiological human characteristics. One might observe that these are the testing grounds for the public's acceptance of big data in human research and, thus, warrant exemplary stewardship.

Yet, the government's persistent use of big data for purposes other than scientific research risks raising questions in the public's mind about the trustworthiness of anybody holding big data about citizens or research participants. Rainie emphasizes that, although the American public expects to exercise control over acquisition and use of personally identifiable information, they also feel increasingly less able to realize their expectation, particularly with reference to commercial and governmental organizations. Unauthorized, excessive and uncontrollable collection and analysis of big data from transactional data, telephone calls, emails or videos does not sit well with many Americans even if searching for terrorists or criminals. Knapp argues, however, that scientists do not simply remain at the mercy of societal currents but, using Lessig's theory of regulators of online activity, suggests how they might actively engage the public in building trust in their big data research activities. Specifically, he suggests that scientists adopt best practices in protecting computerized Big Data (Code), remain completely transparent about their data management practices (Law), make smart choices when deploying digital solutions that place a premium on information protection (Market), and, critically, portray themselves to the public as seriously concerned with protecting the privacy of persons and security of data (Norms). We close section two with Singh asking and offering suggestions for answering a key question: what are the responsibilities of individual citizens for protecting their own privacy while using the Internet as a routine piece of the infrastructure of everyday life. This section of the book sends a clear message: the American public constitutes a major stakeholder in the conduct of research about human affairs using big data.

The third section of the book addresses institutional issues, including the technological, individual and organizational dimensions of the ethical practice of big data research on human affairs. One cannot escape the technological foundations of the concept "big data," particularly its dependence on various types of informatics, computational science, and computer science as well as computer engineering. Thus, the ethical analysis of big data necessarily has a technological component. Indeed, big data in research on human affairs should exemplify what we have called elsewhere the embedding of ethical values in big data technology, specifically guiding the design of big data technology with reference to the ethical values attempting to be fulfilled (Steinmann, Shuster, Collmann, Matei, Tractenberg, FitzGerald, Morgan, and Richardson 2015). The opening chapter of section three (Smart) directly addresses design and implementation of a privacy technology device (a "Black Box"). The Black Box enables analysis of highly sensitive individually identifiable or national security level data without human inspection using algorithms to answer only authorized questions with authorized data from only authorized and identified sources.

Ethical practice in scientific research depends on scientists capable of ethical reasoning about novel situations as they arise in frontier research. Historically, scientific professions have developed and relied upon professional codes of ethics to help their members make sound ethical decisions in their everyday practice. Using the Code of Ethics of the American Statistical Association, Tractenberg offers an approach toward training scientists in the ethical practice of computational science beginning as young scientists and continuing throughout the arc of their careers in a lifelong educational framework called the Mastery Rubric. Yet, ethical practice in human research also depends upon developing strong institutional settings that inculcate, encourage, advise and provide tools for ethical reasoning. In a chapter on designing Data Management Plans for funding proposals, Collmann, FitzGerald, Wu and Kupersmith show the interdependence of the ethical design of projects, ethical practice of scientists and the ethical performance of institutions in big data research. Much current discussion seems intent upon downgrading institutional support for scientists in reasoning about the ethics of big data research on human affairs. In this chapter, we revisit the concept of the 4Rs and highlight the importance of ethical provenance and ethical horizon as scientists draw data from multiple sources and anticipate recurring use of their data years after they completed their own projects.

We close the book with a chapter (Tractenberg) that focuses on scientists' ethical obligations to their own disciplines in the practice of big data research on human affairs. This chapter extends her earlier argument about training with the Mastery Rubric and addresses general issues about the implications of ethical reasoning in becoming a scientist, and, more generally, a thoroughly trained scholar. From this perspective, the practice, inculcation and reproduction over time of ethical reasoning amounts to nothing less than a general responsibility of the scholarly community to foster what she calls "disciplinary stewardship."

Thus, we present a perspective on big data ethical reasoning that is transactional and placed in context. The ethics of big data is a community activity; it is not simply a set of policies, procedures or practices for staying out of the newspaper or checking boxes off funding agencies data management plans. We also locate the design and planning of research projects as a prime site for ethical reasoning because these activities offer opportunities for prospective analysis of potential big data privacy issues and their ethical consequences for research participants, investigators, research organizations and the institution of science. We offer the chapters of this book as illustrations of both specific scientific endeavors such as neuroscience and controversies in our wider society that potentially affect science such as the NSA's surveillance activities. We highlight the importance of, and scientists' responsibility for helping to sustain the American publics' trust and sense of legitimacy of the scientific endeavor. We offer principled approaches, heuristic methods and tools for helping scientists to discharge these responsibilities while planning and implementing their research. Finally, we suggest that national policy should attempt to build the community and capabilities of investigators, research administrators, computer infrastructure staff, and academic leaders for reasoning effectively about the ethics of big data in human subjects research. Assuming we have all the experience we need to dispose of any remaining ethical challenges in big data research about human beings will cost science its most precious asset, the trust and support of the American public.

References

Christakis, N. A., & Fowler, J. H. (2009). *Connected: the surprising power of our social networks and how they shape our lives*. Boston, MA: Little, Brown and Company.

Harris, D. (2013, May 14). We're witnessing the rise of the graph in big data. Retrieved November 12, 2015, from https://gigaom.com/2013/05/14/were-witnessing-the-rise-of-the-graph-in-big-data/

Macy, M. (2015, May). *Big data and the end of science*. Keynote presentation presented at the Northwestern Computational Social Science Summit, Evanston, Ill. Retrieved from http://www.kellogg.northwestern.edu/news-events/conference/csss/2015/agenda.aspx

Nissenbaum, H. (2009). *Privacy in Context: Technology, Policy and the Integrity of Social Life*. Stanford, CA: Stanford University Press.

Steinmann, M., Shuster, J., Collmann, J., Matei, S., Tractenberg, R., FitzGerald, K., et al. (2015). Embedding privacy and ethical values in big data technology. In S.A. Matei, M. Russell, E. Bertino (Eds.), *Transparency on social media—tools, methods and algorithms for mediating online interactions*. New York: Springer.

Wall, M. (2014, March 4). Big Data: Are you ready for blast-off? [News]. Retrieved November 12, 2015, from http://www.bbc.com/news/business-26383058

Part I
Applying a Contextual Analysis of Privacy in Big Data Research

A Theoretical Framework for Ethical Reflection in Big Data Research

Michael Steinmann, Sorin Adam Matei and Jeff Collmann

1 Introduction

Scientific progress has increasingly become reliant on large-scale data collection and analysis methodologies. The same is true for the advanced use of computing in business, government, and other areas. Utilizing massive computational resources, such methodologies can automatically capture and analyze characteristics and processes of entire statistical populations. Sampling is ideally replaced by census-like, complete counting of cases and characteristics. Interconnections between individual elements are turned into graph edges. Complete graphs facilitate research that takes into account case dependencies. They are ideal for detecting diffusion processes in a variety of populations.

When applied to objects of study at micro-scale, such as bio-molecular research, big data research can take instantaneous snapshots of extremely complex systems (genes, proteins, etc.), categorize them, and detect patters or anomalies. When applied at a macro-scale, big data can involve remote sensing networks, including radar, satellites, or telescopes to capture real time information about highly complex phenomena, such as weather patterns, climate change, or mapping the depths of the universe.

M. Steinmann (✉)
College of Arts & Letters, Stevens Institute of Technology, Castle Point on Hudson, Hoboken, NJ 07030, USA
e-mail: msteinma@stevens.edu

S.A. Matei
Brian Lamb School of Communication and Cyber Center, Purdue University, 100 N University Drive, West Lafayette, IN 47907, USA
e-mail: smatei@purdue.edu

J. Collmann
Department of Microbiology and Immunology, Georgetown University, 3700 O St NW, Washington, DC 20057, USA
e-mail: collmanj@georgetown.edu

J. Collmann and S.A. Matei (eds.), *Ethical Reasoning in Big Data*, Computational Social Sciences, DOI 10.1007/978-3-319-28422-4_2

In general, big data is concerned with the exhaustive capture of information about complex systems and the subsequent exploration and explanation by means of an extensive investigation of their elements and characteristics. The depth and comprehensiveness of such an approach are impressive, providing the ability to find the proverbial needle in the haystack. Answers both general and specific can emerge from observing all situations in which even small changes occur. Protein expression in genes is now used as key for opening the door to explaining specific diseases; temperature upticks in the Arctic measured hourly can capture global warming trends; and the propensity of individuals to search for "cold medicine" through Google can be an indicator that the next flu epidemic is about to start. Although critics have pointed out that not all promises of big data research might be realized at the end, the very fact that such promises exist already changes the expectations and attitudes toward research and can lead to all sorts of new experiments with the analysis of data.

The complete, detailed account of phenomena promised by big data represents a boon for research and poses ethical challenges of different kinds. Genetic profiling of entire populations, to take one example, may lead to finding new relationships between genes and disease, and also pinpoint specific individuals who carry such genes. If this is the case, should their genetic potential be reduced or eliminated in the name of public health? If applied to humans this would represent a return to eugenics, a theory and practice long since rejected. If applied to other forms of life, this can lead to a reduction in biological diversity. A search for patterns in open source data collection can be used to inoculate those most vulnerable for a disease before an epidemic has fully developed. But what if searches suggest the emergence of a political movement or protest?

Big data methodologies, thus, have a double potential, both for sharpening our scientific insights and for potentially creating significant ethical dilemmas. At the time being, dilemmas D related to privacy are probably the most important, as they implicate a series of ethical values. However, in the current discussion it seems far from clear what privacy means in each case and how it matters. In addition, due both to its rapidly evolving nature and the hypothetical character that many applications still possess, big data presents a specific challenge to the reflection on ethical issues. A one-size-fits-all approach does not seem appropriate for it. For these reasons, ethical reflection first has to target the various aspects of big data and their impact on human privacy in a differentiated way. We will try to do so by articulating the various normative dimensions that privacy entails. Second, ethical reflection has to address the long-term goals involved in big data research on human subjects. According to the forward-looking character of big data methodologies, our reflection has to be able to anticipate and preempt ethical issues in human subject research by articulating desirable practices and overarching values, such as trust in the integrity of scientific research.

Our methodical approach to ethical reflection on big data can be seen as a well-defined pluralism. This means, on the one hand, that ethical reflection has to address a variety of values, or principles, that cannot be further subsumed under one over-arching value. On the other hand, this ethical pluralism does not mean that in each case conflicting judgments have to be made, which would render ethical reflection practically ineffective. Instead, each situation needs to be judged contextually. Following Nissenbaum (2011), we suggest placing strong emphasis on the context of using big data. Different contexts raise different concerns, which then require analysis with respect to different, specifically chosen ethical principles. This makes it possible both to make meaningful ethical judgments and address the underlying real-world problems in a multifaceted way. In addition, we suggest that special consideration has to be given to the process of decision-making, which often follows a trade-off model, or cost-benefit analysis. Given the dynamic and promising nature of big data, the trade-offs that are made are likely to raise ethical concerns of their own. To minimize such concerns, we propose that each trade-off analysis should not start until a minimum ethical threshold is determined, one under which certain core values should not be traded off for any possible social or individual benefits.

To understand the potential ethical impact of big data, we need to begin by inventorying the essential modalities of big data manipulation that may impact privacy. These are put under what we call the 4R rubric: reuse, repurpose, recombine, or reanalyze, which is detailed below (Sect. 2). We will then start our discussion of ethical challenges with some key definitions related to privacy and its various normative dimensions. These dimensions, born out of core human values, include the principles of non-maleficence, beneficence, justice, autonomy, and trust (Sect. 3). We will continue with a discussion of the contextual nature of privacy and how it motivates and shapes the specific treatment of privacy in each case. To simplify this discussion we propose a "privacy matrix," which helps matching privacy dimensions and given contexts with specific types of trade-offs (Sect. 4). We will propose a heuristic model that uses trade-off analysis, yet overcomes the difficulties typically implied by the utilitarian logic of cost-benefit analyses. We propose a model that starts with determining a minimum "concern threshold" (Sect. 5).

2 The 4R Approach

The ethical impact of big data analysis is born out of two main concerns. On the one hand, big data tends to be exhaustive and precise. Big data deals not with samples, but with populations. Big data is like a sieve, most of the solid matter (objects, people, behaviors, etc.) that goes into it is captured. Only the liquid part of the universe of observation passes through. Also, big data is relational. Being able to count everything, it can determine if the characteristics of the elements under observation are shared and whether the shared characteristics create networks of affiliation or interaction. Elements are understood not only as static objects with certain characteristics, but as generative, evolutionary nodes that can impact or be impacted by other nodes. Ethically, when studying individuals, this means that we can know not only if people are of a certain kind but how susceptible they are to change and from which direction (connection) this change can come.

The other big concern related to big data is born out its ability to be reused, repurposed, recombined, or reanalyzed. This is what we call the 4R challenge of big data. Given the connectedness of big data, its elements can be easily imagined as lego pieces ready to be rearranged and connected to other pieces or collections of pieces (populations) to obtain new insights. The new insights can produce new knowledge but also unforeseen threats for the individuals or population under observation. Since the individuals released their data for a specific use and goal, any further analysis that reveals new processes in that population can lead to insights that were not envisaged or considered desirable by the individuals who contributed the data. For example, a study of Google searches on "pain medicine" to identify arthritic patients that are underserved in specific geographic areas identifies a cluster of searches coming from a college town, where the population is much younger then the typical arthritic sufferer. One can then suppose that the searches are related to pain medication abuse, rather than legitimate use.

Or, suppose that the entire population in a given region, inhabited by a homogeneous native population, is screened genetically to identify the possibility of a relationship between genetic makeup and a given inheritable disease (sickle cell anemia). After the study, the data is also used to identify the genetic haplotypes present in the population. The conclusion is that a significant number of individuals have a genetic inheritance that is not common with that which is considered traditional for that native population. This may affect the notions of cultural identity and nativity.

Such examples, hypothetical only in form, as we will see below, reveal the tremendous ethical difficulties created by big data. In what follows, we will investigate them in more detail, focusing especially on privacy concerns generated by the 4R challenges.

2.1 Reuse

Reuse refers to taking data originally collected for a specific scientific purpose and using them again for comparable purposes in comparable domains. The reuse activities may engage either the original investigators or other investigators. The possibility of reuse, particularly of data originally acquired in scientific activities covered by the Common Rule (Office for Human Research Protection 1993), raises the question of the responsibilities that investigators have for what happens to data once they become available to secondary investigators. Additionally, it poses the question of the responsibilities of secondary investigators for complying with, or reaffirming the conditions of a data set's original collection.

A Case: Reusing genomic data from the Havasupai Indians
Scientists at Arizona State University conducted a series of investigations on blood samples obtained from the Havasupai Indians, a small tribe of people living at the bottom of the Grand Canyon. The studies began when the Havasupai approached ASU for help in understanding the high prevalence of diabetes among their people. In addition to conducting research on the possible existence of a genetic basis for diabetes, ASU scientists re-used blood samples drawn from individual members of

the Havasupai tribe to conduct and publish results on multiple studies of which the Havasupai had no knowledge on a range of other disorders and characteristics. Upon accidentally hearing of the secondary use of the blood samples, the Havasupai sued ASU who, after paying over $1.7 M in legal fees, settled the case by paying $700,000 in reparations and providing other benefits in kind (Harmon 2010; Jacobs et al. 2010).

Ethical implications: The Havasupai claimed harms of honor and cultural integrity, finding specific fault with papers purporting high level of inbreeding among the Havasupai and published claims about their origins at odds with their own traditions.

2.2 *Repurposing*

In contrast to reuse, repurposing refers to taking data originally collected for a specific purpose in a specific domain and analyzing them for unrelated purposes in a domain other than their domain of origin. In addition to the questions posed by reusing data, repurposing big data poses questions about the legitimacy of analyzing data acquired under one privacy context and employing it in a different privacy context.

A Case: Repurposing educational administrative data for scientific research
Social science investigators are finding great value in linking administrative records from multiple administrative data systems and entities to longitudinal datasets at different levels of analysis (individual, family, program, school, etc.). State and local agencies collect the data for multiple purposes, such as program accountability, client tracking, and service effectiveness. "Data" in this case refers to administrative records, established when a person or family applies for social, health or educational services (such as enrolling a child in school). Most states also routinely collect data on newborn babies, their parents (especially mothers), including prenatal care; any birth defects or signs of vulnerable health (e.g., hearing loss registry); and other important social indicators. Public health departments collect a range of data and routinely work with hospitals, clinics and other providers in tracking persons to ensure adequate care and provision of services as well as effective disease monitoring.

Ethical implications: This type of research highlights the importance of considering ethical provenance in employing big data. The scientific investigators have to consider the impact for their own analyses of the conditions under which the data was originally collected, specifically that sometimes the records are used differently from their original purpose. Data that were originally collected without consent for use in research can potentially be repurposed without the knowledge of individuals that the records concern. As the distance grows between the original data source and its eventual uses in research, the gap potentially also grows between what individuals initially expected to happen with their information and the research that might actually be conducted. The NPRM on the Common Rule has recommended that a specific form of consent, Broad Consent, be developed to address these issues but has not yet offered any details. Depending on what actually gets developed, the Broad Consent mechanism might address the question of ethical provenance because it enables notice; but to be effective, it should offer an opportunity to shield one's records at any point in the process from initial data acquisition to incorporation in a research project.

2.3 Recombining

The term big data frequently evokes a process of combining and recombining data from various sources to achieve greater analytic yield. In addition to the questions posed by reusing and repurposing data, recombining data poses questions about the possibility of developing new information not available to the investigator simply from the constituent data sets. From a privacy perspective, recombining data potentially enables re-identification of individuals from data that contains no specific identifiers or has been intentionally stripped of identifiers. Indeed, a research project may posit such re-identification as an explicit goal, for example in attempting to track persons with a infectious disease across time and space. The possibility of re-identification through recombining data raises questions about privacy protection distinct from information protection.

A Case: Recombining data to forecast forced migration
Investigators at Georgetown University are recombining multiple sources of big data about forced migration to better forecast, respond to, and help alleviate the consequences of humanitarian crises. Because detailed local data is difficult to obtain in a timely manner, this project explores the effectiveness of using open-source, online data to help identify indirect indicators of displacement/forced migration. Indicators relevant to this project include: economic, political, social, demographic and environmental changes affecting movements; intervening factors such as government refugee policies; and community and household characteristics. Parsing irrelevant information from the true indicators, calibrating results, understanding how these indicators change through time, and identifying and removing potential bias, requires large-scale data analysis and potentially, new computational methods for developing meaningful descriptive and predictive models. To date, the big data Georgetown uses for this study include open-source media articles and Twitter data. The investigators have access to EOS, a vast unstructured archive of over 700 million publicly available open-source media articles that has been actively compiling since 2006. New articles are added at the rate of approximately 300,000 per day by automated scraping of over 22,000 Internet sources in 46 languages across the globe. The project also collects data from Twitter—hundreds of thousands of tweets per day for the last 6 months. When relevant, the investigators also draw data from the scholarly literature of history, anthropology,

economics and other social sciences as well as the gray literature of governmental and non-governmental organizations. Long term plans include adding data from the archives of collaborating international and non-governmental organizations.

Ethical implications: This case highlights the importance of protecting the results of big data analyses with the explicit intent of better describing and aiding specific individuals or communities; that is, potentially re-identifying people and, thus, creating privacy breaches for the purpose of humanitarian aid.

2.4 *Reanalysis*

Big data archives have been assembled, particularly in public health and healthcare, with comparative or longitudinal purposes in mind. Although investigators may identify some specific objectives at the time of the archives' creation, they also expect and hope that new uses may emerge as scientific knowledge grows, lines of inquiry develop and techniques for extracting new information from collected data sources become more sophisticated.

A Case: Reanalyzing Newborn Screening Data
State mandated programs provide screening and data collection for 4 million newborns in the U.S. each year. After newborn screening is completed, the residual dried blood spots (RDBS) and data can be stored for quality assurance and research purposes depending on state practice and statutes. Currently, fourteen states store RDBS for research purposes. Storage of RDBS and data can range from a few months to the entire life of the program depending on the state. For example, California has stored the RDBS and data for newborns born in California for the past 52 years. Programs in Minnesota and Texas lost law suits alleging use of their RDBS for purposes not included in the original parental consent, including but not limited to research. Indiana is currently involved in an active lawsuit.

The State of Minnesota was sued by 21 families who alleged the program's collection, use, storage, and dissemination of RDBS and test results without written, parental consent violated the Minnesota State Genetic Information Act of 2006. Some of the dissemination of the RDBS and data associated with newborns were for research purposes. First begun in 2009, the lawsuit was initially dismissed in district court and the dismissal was upheld on appeal. However, on Nov. 16, 2011, the Minnesota Supreme Court ruled that the use of the blood spots and test results for anything other than the initial screening was not explicitly authorized in statute. The State of Minnesota settled the lawsuit and destroyed 1.1 million RDBS and their associated data prior to the November 16, 2011 ruling. The Minnesota legislature revised the statutory language explicitly to authorize short-term storage and the use of blood spots and test results for program operations, and to require written, informed consent for long-term storage and use of blood spots or test results.

Ethical implications: This case highlights the importance of the concept of ethical horizon and the conditions for building trust in subjects who understand that their data may be used in ways not yet imagined. Parents provided information about their children with no knowledge of its potential use outside that context, including scientific research. Like the Havasupai case, when public health officials or scientists take actions beyond the terms of original consent for data collection, they jeopardize the public's trust in the institution of science. In contrast to the Havasupai, the case of newborn screening potentially affects a majority of families in the United States and a major resource for public health and genetic research for years to come.

3 Pluralism of Principles

The ethical concerns that are raised by the cases mentioned above deal for the most part with privacy concerns. Given the complexity of the cases, privacy is more than the proverbial "right to be left alone" as defined by judge Brandeis (Olmstead vs. United States 1928). In fact, although there is no doubt that privacy can be seen as a value, or ethical principle, it is hard to define its normative dimension specifically, compared to other values, and to show what exactly its practical implications are. In the following, the terms "value" and "principle" can be used interchangeably, although we will mostly use the latter in referring to privacy, as it seems more directly action-related than the term "value." Still, ideas or states of affairs are "values" not in and for themselves but because they entail specific actions. The meaning of the term "value" can be defined as a quality that makes the corresponding actions desirable or obligatory. If privacy is called a "value," then, certain rules should also be called up to regulate certain actions so that specific human qualities are protected.

We start from the assumption that privacy alone cannot be defined as value. Although it is possible to say: "You should not do x because it violates privacy," such a statement is incomplete. One can always ask: "But why does privacy matter?" This means that privacy is embedded in a set of other values. There are five values, or principles that we have identified in our previous work as being associated with privacy. These principles also have clear and direct utility in designating practical guidance for protecting privacy in research contexts. They are: nonmaleficence; beneficence; justice; autonomy; and trust. Taken together, the five principles define our approach as a well-defined pluralism. We understand the principles in the following way.

Do no harm or non-maleficence: harm has to be widely conceived in the case of big data. While physical harm is not likely to occur, or only as a remote consequence, a violation of privacy can have measurable effects, for example in the case of a financial loss. Psychological distress also has to be counted as harm and is measurable to a certain degree. The well-known formula that is used in medical ethics—"primum non nocere" ("first, no harm")—can be applied to the use of data, too. The question whether individuals are harmed or not has primacy over other concerns. If real harm occurs, other normative principles, such as autonomy and trust, become secondary. In a sense, all normative questions concerning big data have to do with some sort of harm, but only impacts that can be seen as a painful experience of some sort should be subsumed under the category of harm. For example, a certain violation of autonomy can be unwanted and undesirable without though producing a directly measurable negative effect. (This does not mean that in such cases violations of autonomy bear no normative concern, it just means that 'harm' might not be the most appropriate category to articulate these concerns.) It has to be noted that a certain amount of harm can be seen as ethically permissible. In medical ethics, the category of minimal risk is used (Office for Human Research Protection 1993). It can also be applied to research projects in big data, provided that minimal risk can be defined accurately and responsibly in this field.

Do good or beneficence: beneficence can be defined as concern for the well-being of others. Compared to non-maleficence, beneficence can often be seen as supererogatory (see Beauchamp 2013). Research projects can be permissible even if they do not maximize the well-being of participants or the public, at least not directly, and researchers have no obligation to contribute to the increase of the well-being of others. On the other hand, they are almost always obliged to avoid harm. It has to be noted that beneficence is not necessarily identical, or reducible, to an interest in promoting economic benefits but can be used in a wider sense (even if cost-saving and similar outcomes can certainly be seen as beneficial). Beneficence can play a crucial role in trade-off analyses (see below). However, especially with regard to beneficence the pluralism of our approach comes to bear. Having intentions of doing good can only be used as a legitimizing principle of actions if the other normative principles mentioned here are equally addressed. As a rule of thumb, it seems appropriate to use the principle primarily in a critical way, by asking, for example, whether a certain use of data does indeed yield any genuine benefits to participants or the public. In general, beneficence has to be applied with some caution, as big data professionals sometimes tend to exaggerate the social benefits of their innovations. In such cases, the burden of proof lies with those who claim to act according to an interest in beneficence, not with the users or recipients of the data applications that they create.

Justice: the principle is concerned whenever opportunities, rights, and goods are to be distributed among the individuals or groups who have been targeted by big data analyses. Violations of justice are social disadvantages of all kind, or acts of discrimination (see Executive Office of the President 2014). For example, an analysis of big data can lead to the result that certain socio-economic groups can be treated differently compared to others because they are less likely to benefit from opportunities that are given to them. Some evidence exists that big data are used to sort out less lucrative social groups (Marwick 2014). However, discrimination can also occur when data are *not* used to examine the existing disadvantages of groups. In such cases, the principle of justice can be critically applied to the design of research projects, not just to the use of their results.

Autonomy: the principle can be applied both in a concrete and a general sense. In the concrete sense of the term, autonomy is a well-established principle in other fields of applied ethics, where it refers to the freedom and capability of decision-making in individuals. The procedure of informed consent, which is both an ethical and a legal requirement in research projects, is meant to ensure that all relevant information about a given study is disclosed to the participants and the latter have been afforded the opportunity to deny or modify their participation (for recent discussions, see White 2013). One of the questions that is addressed in other parts of this book (Collmann, FitzGerald, Wu and Kupersmith) asks to what degree the tool of seeking informed consent is realistic in big data studies. In the general sense of the term, autonomy refers to the social, political and economic practices of individuals that allow them to realize freedom (see Cohen 2012). For example, citizens have to be given the opportunity to articulate their political opinion freely, while customers are to be given the opportunity to choose goods according to their liking and establish contracts of all kinds with other economic subjects. In this sense, autonomy is an overall quality of social practices that is not reducible to isolated acts of decision-making. Democratic societies establish such practices as open opportunities for citizens to exert and cultivate autonomy, for example through equality before the law and electoral procedures. These opportunities can be seen as desirable goal within a society. Although their violation does not necessarily cause any direct harm to individuals (see above), it might restrict the overall autonomy they have, or perceive to have, at their disposal. Big data has a potential to undermine practices of autonomy, for example through the possibility of tracking and profiling individual behavior at all times, or through the use of data for the prediction of individual decision-making.

Trust: this principle refers to the informal agreements that have to exist among the members of society in order to allow individuals to pursue their personal good. It also has to exist between individual members of society and their institutions. Trust enables individuals to engage in innovative social projects and take risks. It eliminates the burden of securing the appropriate conditions each time that an individual acts. Trust is closely linked to autonomy insofar as individuals have to have confidence that their autonomy is both realizable and protected. Trust, however, has also to exist in cases where there is an asymmetry between the members of society. For example, parents have to trust schools to treat the data of their children confidentially, while schools have a right, and perhaps an obligation, to collect all kinds of sensitive data for the purpose of education. Like beneficence, trust should only be used as a legitimizing principle if the other normative principles are also addressed. For example, trust often has to be balanced with autonomy, and such trade-offs have to be made in a transparent and revisable way. Again like beneficence, it seems best to use the principle primarily in a critical way, for example, in order to assess whether the use of personal data might erode trust in the long run, or whether there is a discrepancy between the trust that is requested from the participants of studies and the way trustworthiness is established by the respective institutions and research personnel.

This overview of the five principle documents the need for a well-defined pluralism. No single principle is sufficient to address all concerns that are raised by the use of big data. At the same time, the principles are not introduced arbitrarily but complement each other by addressing aspects that each principle, taking in an isolated manner, has to leave out. But the pluralism also allows us to show the importance of trade-off analyses, as there can be cases in which the principles do not so much complement each other but compete. When this happens, the relevance of the respective normative points of view has to be addressed, which then makes it necessary to address the threshold conditions that exist for each principle's point of view. In the following, we will say more about this point. It can also be noted that similar attempts in the data science community to establish ethical principles have led to a comparable set of principles. See the Menlo Report (Dittrich and Kenneally 2012).

The pluralism of principles allows us to address in a nuanced way the practices that are involved in the use of big data. Non-maleficence, beneficence, and justice have in common that they can be seen as action-related principles. Each concerns the legitimacy of the actions that are performed, or of the consequences that directly follow actions. On the contrary, autonomy and trust are agent-related. They concern the attitudes and perceptions of individual agents insofar as they are dispositions for an infinite variety of actions. Action-related principles concern individuals as targets, or objects, of actions. Individuals are referred to insofar as they might be harmed, benefited, or disadvantaged. Agent-related principles, in turn, concern individuals as spontaneous agents, or subjects, by asking what is necessary for them in order to maintain their agency. In simpler terms, action-related principles are concerned with the protection of subjects, while agent-related principles aim at empowering them.

For a full assessment of the privacy concerns raised by big data, it is necessary to keep this distinction in mind. The individuals that are targeted by big data uses are regular members of society who pursue active practices on the various levels of social life. It seems easy to confuse data related to individual agents with mere data points and to assume that individuals are helped if only the flow of data is optimized. The idea that one can optimize human life by optimizing data technologies has rightfully been called "solutionism" (Morozov 2014). But insofar as big data analyses can interfere with the realm of individual agency and potentially pervade all aspects of social life, one has to consider all aspects that are relevant for the full realization of democratic practices. This makes it necessary to establish feedback loops that go beyond the realm of "pure" data analysis and involve the real-world concerns and needs of individual agents.

In more practical terms, the distinction between individuals as targets and as agents has been addressed by the difference between the "restricted access theory" and the "control theory" (see the overview in Tavani 2008). For these theories, privacy is realized either by restricting the access to personal data or by giving data sources control over the data they want to share. Both theories can be used legitimately, as individuals and groups both have to be protected passively and need to be involved actively in the data collection in case the possibility for such an involvement exists.

At the same time, the distinction between the principles outlined here allows us to distinguish immediate from long-term outcomes. For example, if predictive techniques can be implemented effectively they are likely to change the fundamental conditions of individual agency. In such cases, no immediately measurable outcomes have to be identified in order to raise concerns. A change in the basic conditions of social practices may call for ethical reflection even if no immediate harm can be identified at present. In addition, big data can concern both individuals and communities. While harm, seen as measurable impact, is more likely to be identified in individuals, the interests of communities can also be addressed using the principle of justice.

4 The Variety of Contexts: The "Privacy Matrix"

Besides the variety of normative principles, it is pertinent to consider the practical context in which privacy matters. Given the fact that big data is generated in a broad variety of human affairs, sometimes in an automatic way and under some assumptions of publicity (e.g., geo-tagged photos posted on the web, or social media links and posts announcing political attitudes), while at other times with the expectation that the information will be strictly guarded (genetic profiling), it is imperative to consider the impact of context (social, scientific, economic, political) on the privacy regime of each situation.

A very similar approach has already been suggested by Nissenbaum (2011). According to her, "we must articulate a backdrop of context-specific substantive norms that constrain what information websites can collect, with whom they can share it, and under what conditions it can be shared" (32). In developing her approach, Nissenbaum warns that we need to take into account all possible contexts, not just commercial ones. Only by using a contextual approach can we navigate the complex decision-making landscape of privacy. Heuristically, she advises that we "locate contexts, explicate entrenched, informational norms, identify disruptive flows, and evaluate these flows against norms based on general ethical and political principles as well" (38).

Given the dynamic nature of big data, as it is evident in the uses described in Sect. 2, the relation to context seems to become ever more relevant. If big data are used according to their potential, they are very likely to switch contexts, for example between research and commercial applications, or between commercial applications and the government.

To simplify this discussion we have proposed elsewhere a "privacy matrix," which helps matching privacy dimensions in given contexts that involve interactions between data collection agencies, scholars, commercial agents, political actors, and ordinary individuals (see Steinmann et al. 2015).

The privacy matrix				
Specifying principles	Privacy contexts			
	Social	Government	Commerce	Science
Nonmaleficence				
Beneficence				
Justice				
Autonomy				
Trust				

The matrix is structured in two dimensions: contexts and normative concerns. The normative concerns are based on the five principles mentioned above: non-maleficence, beneficence, justice, autonomy, and trust. The contexts are broadly defined, as social, government, science, and commercial. "Social" refers to open, public domains outside of business entities, research institution, or government bodies. While it is unlikely that genuine research of big data can be conducted in this realm, data still occur and research tools might get used, in one way or the other. The main idea of this privacy matrix is that the same ethical concerns can become more or less sensitive or more or less tractable according to each context, since in each context the amount of disclosure, the nature of disclosure, and the ultimate effects of disclosure vary in gravity. Also, the legal and moral expectations, especially as encapsulated in norms and legislations, allow transactions to be more or less permissive when it comes to privacy.

The matrix is to be used as a heuristic tool. Given a certain situation or research project that involves big data in one of the columns (that is, contexts), one walks down the value or concern list, considering at each step the specific nature of normative implications for the given context. Naturally, it is desirable that all ethical concerns should be considered important, since we cannot make capricious decisions as to what matters or not. Yet, not all principles are equally applicable. Ultimately, the heuristic process demands that we zoom in on the most relevant normative concern for a given context and try to answer the question: how will privacy be protected in such a way that the relevant concerns are met, while the data is still usable?

The matrix does not presuppose that data belong essentially only to one sector. By its nature, data cross borders and can be linked in a multitude of ways. As already said, data can "travel" from closed settings in, say, educational institutions to commercial institutions and the government. For example, data generated from individual practices, such as movement in space, can become relevant in commercial and governmental settings. Big data is protean in nature insofar as use and impact cannot be stated definitively by focusing only on one specific context or one specific way in which data are presently used. On the other hand, while data are likely to "travel" in these ways, their respective use is still always relevant in one specific context, for specific users and their purposes. The multi-contextuality of big data does not dispense the ethical reflection from considering each context specifically.

The rubrics in the matrix are thus not separate universes that endow data with an essence that prevents them from being used in other rubrics. They rather have to be seen as areas of use and relevance. The basic meaning of the matrix is in this sense dynamic, not static. The underlying normative idea is to determine specifically who controls the access and exploitation of data in each case. If data, for example, are used and stored within a governmental context, the task is to determine in which way the respective agencies can use the potentials of data sets. And, if data "travel" across the borders of contexts, how much of the control previously established will be lost, and which new purposes will be added?

Taking up Nissenbaum's (2011) use of the term "appropriateness," we can say that the purpose of the matrix is to define the appropriate concerns in each case of the use of data. On the side of subjects, or data sources, it is important to consider their rights and legitimate expectations. On the side of users, the legal and institutional responsibilities have to be established. In addition, all these considerations have to be qualified as revisable given the protean nature of big data.

It is worth noting that the contexts might also relate to different strategies of inclusion and exclusion. While there is a tendency to understand privacy predominantly as a way of restricting access to personal data, harm can also result from not considering some private data. If big data research leaves out, say, vulnerabilities due to race or social status, then the protection of privacy, paradoxically as it seems, becomes harmful. For this reason, it helps keeping both principles of non-maleficence and beneficence in mind and ask for each case: have we done as little harm as possible, and have we done as much good as we can?

5 Trade-off Analysis and Threshold Conditions

The matrix can also be employed in a heuristic methodology that uses a modified version of trade-off analysis, which may overcome some ethical difficulties typically implied by the utilitarian logic of trade-offs, such as those mentioned by Kelman (1981) and discussed in detail in the literature (Palm and Hansson 2006; Elgesem 2002).

We suggest a model that starts with determining a minimum "concern threshold" for each dimension. In other words, our model starts with the assumption that trade-off analysis needs to take into account that the results of a cost-benefits analysis cannot lead to reducing any of the dimensions (autonomy, trust, etc.) beyond a minimum acceptable value (threshold), which in no case can be 0. The "normal" minimum thresholds are to be determined as much as possible in absolute terms, regardless of context. Ideally, there should be a minimum of universally applicable level of beneficence, harm avoidance, trust-protective behavior, or autonomy-defensive procedures for all contexts. If this cannot be determined as an absolute value, it should, at the very least, be determined within a narrow band of variation.

For example, for all big data collection that involves harvesting information from social media, autonomy should be protected across contexts in such a way that in none of them the fundamental right to free expression is reduced to zero. In other words, in no context should the data collected be utilized in a manner that diminishes the right of the individuals observed to decide on what to believe or say or that leads to retaliatory measures against them by state or non-state actors. Of course, above this threshold some projects can disclose more and some less about what was said in what context by what type of user, according to the nature of the data collection process. If the data was collected, for example, from a large, government-sponsored organization using social media (e.g., a health peer-support forum for former military personnel suffering of PTSD), anonymization measures need to be strict even if the communication was made in public, under a user's own name. On the other hand, if data was collected from a publicly available site, say, Twitter or Wikipedia, some information about the users can be used in the project, since such material is comparable, in terms of publication privacy, to letters to the editor or other publicly made statements. Yet, again, some publicly available sites or platforms, are public only in that anyone can sign up or apply to become a member. If upon joining the sites or platforms researchers enter spaces defined as private or "closed" the natural expectation of the members that the information is to a certain extent private should not be violated. Furthermore, any participation should be accompanied by appropriated disclaimers, announcing that researchers act in a research context and their goal is to collect data for research purposes only.

To make things more complicated, social media data collection can at times be automated. Specialized software can be instructed to "spider-walk" and harvest information from a variety of social media groups through tools and procedures provided by the platform and site administrators. Such tools, typically called APIs (Application Programming Interfaces) reveal and make available text, images, likes, or comments posted by site users, including those acting behind the "firewall" of "private groups." These applications should not be hidden behind the veil of technological automation. The control that the members imposed on access demands higher level of privacy protection, which demands informed consent from all members concerned.

Furthermore, trade-off analysis needs to include transparency as a core procedural assumption. In other words, the terms of the elements that are traded against each other, their measure and significance, and the exact cost incurred for each benefit should be visible and actively propagated. In addition, transparency needs to be prospective, not retrospective. The terms of the analysis and the presumptive results need to be announced before, not after, the trade-off process is completed.

Finally, transparency should aim at generating community participation in the trade-off process (Milne 2000; Hann et al. 2002). Transparency without active input from all sides is deceiving and in no way conducive to ethical behavior, quite to the contrary. Participation can be achieved in various ways and presents specific challenges within each community. These challenges can partly be practical, because community engagement can be difficult and cost-intensive, and partly arise from the context of data use. Commercial entities, for example, have a certain right to keep their practices secret if only to allow them to stay competitive, while in research it is often not possible to disclose all purposes of a study to its participants without distorting the results. Still, it is important to uphold the obligation that data research has to be transparent, or at least as transparent as possible given all legitimate considerations, to the subjects who provide the data in the first place. Feedback loops have to be created that involve the data sources, that is, individual agents, with the opportunity to exert their active agency.

Transparency is not the only moderating factor in determining the ethical impact of big data collection and analysis. Another factor that needs to be taken into account is specificity with respect to context. Context, as explained above, is to be regulated in each case by a specific procedural approach that emphasizes the particular expectations of privacy of data providers (individuals) and data collectors (e.g., researchers). For any ethically responsible trade-off analysis, maximum attention has to be paid to these requirements. In a commercial context, for example, transparency requires the user to be informed fully and in detail how data is being collected, what is done or will be done with it, and whether any possible sunset policies exist. Methods of redress and opt-out need to be offered. However, since the transaction is conducted on a commercial basis, involving monetary exchanges or fiduciary interests, which require tracking down payments and material interests, as well as adjudication of ownership of content or other type of intellectual property, perfect anonymity cannot be enforced. Also, disclosure to third parties, such as governmental agencies, in cases involving criminal acts (drug dealing or sexual exploitation on an open social media site, material support to terrorist organizations through fundraising, etc., infringement of intellectual property, abuse or violence) should be considered legitimate types of disclosure. In a research context, on the other hand, transparency should include more than informing the user as to the methods to be used to protect identity but also reassure her that her identity will not be disclosed even in situations that are currently considered within the purview of criminal law. Researchers that operate under terms of use and privacy statements regarding data collection that emphasize the absolute anonymity of the respondents should makes sure that they use all necessary means to guarantee it and to inform the users on the measures they will use to do so, even in situations that would typically force data holders to release personally identifiable information. Furthermore, researchers need to pro-actively enforce procedures of anonymization by data aggregation or reduction of data granularity that avoids disclosure of private data through re-analysis and recombination.

6 Conclusions

The elements that are mentioned in this article—the well-defined pluralism of normative principles, the matrix listing the various privacy contexts, and the challenges for any trade-off analysis to consider minimum threshold conditions—exemplify the specific challenges that arise through the new methods of big data analysis. Ethical reflection needs to be adaptive to the evolving nature of big data. At the same time, it has to develop conceptual tools that can be fine-tuned to the various cases that can arise. Some concerns mentioned in this article are still hypothetical and will perhaps need to be changed once the results, failures or successes, of big analysis become evident in the future. On the other hand, while this means that the ethical reflection is to certain parts also still hypothetical, it affords at least the possibility of engaging the community of researchers and data professionals from the onset of new developments.

References

Beauchamp, T. (2013). The Principle of beneficence in applied ethics. In E. N. Zalta (Ed.), *The Stanford encyclopedia of philosophy*. http://plato.stanford.edu/archives/win2013/entries/principle-beneficence, Accessed 13 November 2015.

Cohen, J. (2012). Configuring the networked citizen. In A. Sarat, L. Douglas, & M. M. Umphrey (Eds.), *Imagining new legalities: Privacy and its possibilities in the 21st Century* (pp. 129–153). Stanford: Stanford University Press.

Dittrich, D., & Kenneally, E. (2012). *The Menlo report: Ethical principles guiding information and communication technology research*. US Department of Homeland Security. http://www.dhs.gov/sites/default/files/publications/CSD-MenloPrinciplesCORE-20120803.pdf, Accessed 13 Nov 2015.

Elgesem, D. (2002). What is special about the ethical issues in online research? *Ethics and Information Technology, 4*(3), 195–203.

Executive Office of the President. (2014). *Big data: Seizing opportunities preserving values*. http://www.whitehouse.gov/sites/default/files/docs/big_data_privacy_report_may_1_2014.pdf, Accessed 13 Nov 2015.

Hann, I.-H., Hui, K.-L., Lee, T., & Png, I. (2002). Online information privacy: Measuring the cost-benefit trade-off. *ICIS 2002 Proceedings*, 1.

Harmon, A. (2010). Indian tribe wins fight to limit research of its DNA. *The New York Times*. Retrieved from http://www.nytimes.com/2010/04/22/us/22dna.html

Jacobs, B., Roffenbender, J., Collmann, J. Cherry, K., LeManuel, L., & Bassett, K., et al. (2010). Bridging the gap between genomic scientists and indigenous peoples. *Journal of Law, Medicine and Ethics, 38*(3), 684–696. http://arep.med.harvard.edu/pdf/Jacobs-JLME_10.pdf

Kelman, S. (1981). Cost-benefit analysis: An ethical critique. *Regulation, 5*, 33.

Marwick, A. (2014). How your data are being deeply mined. *New York Review of Books*. January 9. http://www.nybooks.com/articles/archives/2014/jan/09/how-your-data-are-being-deeply-mined/?pagination=false, Accessed 13 Nov 2015.

Milne, G. R. (2000). Privacy and ethical issues in database/interactive marketing and public policy: A research framework and overview of the special issue. *Journal of Public Policy & Marketing, 19*(1), 1–6.

Morozov, E. (2014). *To save everything, click here: The folly of technological solutionism*. New York: Public Affairs.

Nissenbaum, H. (2011). A contextual approach to privacy online. *Daedalus—The Journal of the American Academy of Arts & Sciences*, (Fall), 32–48.

Office for Human Research Protection. (1993). *Protecting human research subjects: Institutional review board guidebook*. Washington, DC, U.S. Government Printing Office. http://www.hhs.gov/ohrp/archive/irb/irb_guidebook.htm, Accessed 13 November 2015.

Olmstead v. United States. (1928). 277 U.S. 438, 48 S. Ct. 564, 72 L. Ed. 944.

Palm, E., & Hansson, S. O. (2006). The case for ethical technology assessment (eTA). *Technological Forecasting and Social Change, 73*(5), 543–558.

Steinmann, M., Shuster, J., Collmann, J., Matei, S. A., Tractenberg, R. E., & FitzGerald, K., et al. (2015). Embedding privacy and ethical values in big data technology. In S. A. Matei, M. G. Russell, & E. Bertino (Eds.), *Transparency in Social Media* (pp. 277–301). Springer International Publishing. Retrieved from http://link.springer.com/chapter/10.1007/978-3-319-18552-1_15

Tavani, H. (2008). Informational privacy: Concepts, theories, and controversies. In K. E. Himma & H. Tavani (Eds.), *The Handbook of Information and Computer Ethics* (pp. 131–164). Hoboken: Wiley.

White, L. (2013). Understanding the relationship between autonomy and informed consent: A response to Taylor. *The Journal of Value Inquiry, 47*(4), 483–491.

Part II
Ethical Reasoning Beyond Privacy in Big Data

Part II
Ethical Reasoning Beyond Privacy in Big Data

The Privacy Preferences of Americans

Lee Rainie

The promises of "big data" seem nearly boundless. Among them: new efficiencies and conveniences in daily life and economic activities; predictive modeling that will helpfully meet human needs; deeper self-awareness and beneficial behavior change as people "quantify" their lives; fuller understanding of people's social interactions; extensive analytics that can make research of all kinds more insightful (especially in the domains of health and wellness); vastly more inputs that give richer pictures about the quality of the environment; "smarter" communities, homes, and workplaces that are safer, cleaner, and cheaper; and more "transparent" institutions that are responsive to their stakeholders.

At the same time, the perils of big data are also clarifying. The collection and analysis of all this data pose threats to people's privacy; potentially increase social and political divisions as some groups find it easy to find ways to exploit the advantages of big data and others struggle to navigate a data-saturated world; possibly cut deeply into fundamental human agency as people find their choices proscribed by algorithms that are applied to the datasets; and conceivably overwhelm public institutions and community social systems in their capacity to set rules and form norms around how the data are used.

The promise and peril of big data frame the dilemma of the networked age, as people function in loose, far-flung personal networks. There are considerable incentives in such a world to disclose and share a great deal of information. Doing so helps people deepen friendships, form communities, become more successful economic agents, and accomplish their goals. Sharing information about oneself and soliciting material from others potentially helps networked individuals realize who can help them when they have problems to solve or decisions to make. In a world of networked individuals, the balance sheet of calculations people make

L. Rainie (✉)
Internet, Science, and Technology Research, Pew Research Center, 1615 L Street, NW, Suite 800, 20026 Washington, DC, USA
e-mail: Lrainie@pewresearch.org

J. Collmann and S.A. Matei (eds.), *Ethical Reasoning in Big Data*,
Computational Social Sciences, DOI 10.1007/978-3-319-28422-4_3

about disclosure has changed from prior eras that were more characterized by close, tight-knit social units. They know there are benefits to personal disclosure. And it takes more effort and calculation to remain masked and hidden.

While most civilians do not know they are living at the dawn of the big data era, they surely have expressed their views over the years about the value they place on personal privacy and the ways in which they act when they are asked to share personal information. The long-standing research on Americans and privacy illuminate the degree to which people feel there should be limits on the collection of personal information and the ways in which companies and the government should behave once their personal data have been collected. Moreover, this privacy research can help the architects of big data research create ethical rules and methodological schemes for using big data in ways that people will accept and might willingly embrace in their lives. The alternative is that the research community can ignore Americans attitudes and behaviors around privacy. That would surely deepen public cynicism and distrust, and, possibly, people's willingness to share their information in ways that could produce societal benefits.

The Pew Research Center has conducted surveys and extensive focus groups and interviews around privacy and disclosure issues since 2000 (Pew Research 1 2000). Over that period, it has gained six fundamental insights about the attitudes and behaviors of American adults that could be applied to the collection and analysis of big data.

1. **The balance of forces has shifted in the networked age. People are now "public by default and private by effort," in the words of communications scholar danah boyd.**[1]

Americans have a strong sense that many entities are gathering information about them. In the internet era, data gathering is a persistent and pervasive practice. People have long been familiar with the idea that they under *sur-veillance* regimes where important and powerful organizations are monitoring them. That process continues with new fervor today as corporations, law enforcement agencies, and government intelligence analysis marches on. At the same time, new forces are unfolding in digital times. After the emergence of social media like email, blogs, Facebook, and Twitter, ordinary citizens are increasingly aware that they themselves have the capacity to monitor the activities of those more powerful. This could be called *sous-veillance* and it underlies many of the efforts to make major organizations more open and accountable. Finally, people know peer-to-peer *co-veillance* allows them to chronicle the environment around them, including those in their vicinity or whose activities are posted in social media newsfeeds.

This systemic monitoring and documentation feels like a "part of everyday life—neither sinister, nor benign. It's the way things are and most of the time it doesn't occur to me to think about it," said one online focus group participant in a Pew Research privacy study (Pew Research 7 2015). Indeed, large numbers of internet

[1]Boyd (2014).

users (at this point, 87 % of the adult U.S. population) know that key pieces of their personal information are available about them online, ranging from photos of them to their political views and affiliations. At the same time, growing numbers of internet users (50 %) say they are worried about the amount of personal information about them that is online—a figure that has jumped from 33 % who expressed such worry in 2009 (Chart 1).

Still, even as they recognize that a lot of information is collected about them, Americans are anxious in basic ways about what this means about their privacy. First, they lack confidence that they have control over their personal information: 91 % of adults in the 2014 Pew Research Center survey "agreed" or "strongly agreed" that consumers had lost control over how personal information was collected and used by companies (Pew Research 6 2015). Second, they express a consistent lack of confidence about the security of everyday communications channels—particularly when it comes to the use of online tools (Pew Research 3 2014). For example:

- 68 % feel insecure using chat or instant messages to share private information.
- 58 % feel insecure sending private info via text messages.
- 57 % feel insecure sending private information via email.

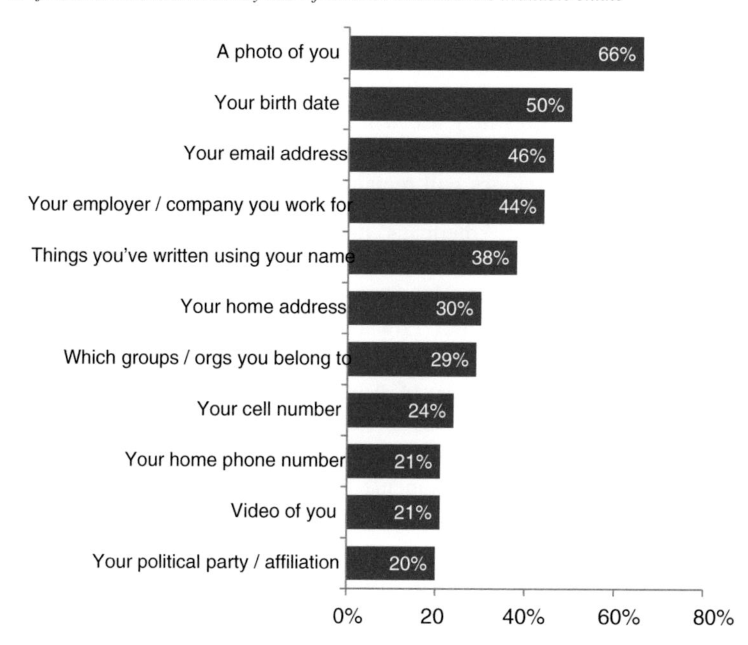

Chart 1 Personal information online. *Source* Pew Research Center survey July 11–14, 2013

- 46 % feel "not very" or "not at all secure" calling on their cell phone when they want to share private information.
- 31 % feel "not very" or "not at all secure" using a landline phone when they want to share private information.

Third, they exhibited a deep lack of faith in organizations of all kinds (public or private) in protecting the personal information they collect. Only tiny minorities say they are "very confident" that the records maintained by these organizations will remain private and secure (Pew Research 6 2015).

- Just 6 % of adults say they are "very confident" that **government agencies** can keep their records private and secure, while another 25 % say they are "somewhat confident."
- Only 6 % of respondents say they are "very confident" that **landline telephone companies** will be able to protect their data and 25 % say they are "somewhat confident" that the records of their activities will remain private and secure.
- **Credit card companies** appear to instill a marginally higher level of confidence; 9 % say they are "very confident" and 29 % say they are "somewhat confident" that their data will stay private and secure.

Online service providers are among the least trusted entities when it comes to keeping information private and secure (Pew Research 6 2015):

- 76 % of adults say they are "not too confident" or "not at all confident" that records of their activity maintained by the **online advertisers** who place ads on the websites they visit will remain private and secure.
- 69 % of adults say they are not confident that records of their activity maintained by the **social media sites** they use will remain private and secure.
- 66 % of adults say they are not confident that records of their activity maintained by **search engine providers** will remain private and secure.

Implications for big data: Americans' distrust in the organizations in charge of protecting communications emerges in the same season as several major corporations announced that key personal information about account holders had been breached and several months after Edward Snowden, a contract worker for the National Security Agency (NSA), leaked information to international news media about widespread NSA surveillance of Americans' phone and email records. It is the period in which public attitudes about privacy issues demonstrated a more urgent tone than in previous years. Those hoping to use big data would be wise to make sure that data-sharing arrangements they have with other organizations are secure and that there be mechanisms to disclose clearly the ways in which the data will be used and who will have access to it. Moreover, Americans would be comforted to know if there were data breaches or successful efforts to use the data to re-identify participants. They would also appreciate a process to gain redress from harms caused by data breaches or re-identification efforts.

2. **Privacy is not binary—either on or off—for most Americans. The context and conditions of information transactions matters**.

People's decisions about whether to disclose information about themselves and how to disclose the data are highly context dependent. It depends on what personal information is at issue; who is watching or capturing the data; and what the "value proposition" for personal disclosure. A stark affirmation of this was evident in a 2015 Pew Research Center survey that posed several possible scenarios with Americans and asking whether they would accept the tradeoff of sharing personal information for a good or service (Pew Research 7 2015). The survey covered six possible scenarios and the overwhelming majority of adults—83 %—were open to at least one information-sharing scenario. But only 4 % were open to every scenario; in other words, their answers to whether they liked these information transactions were "it depends." Two examples from the survey illustrate this.

One scenario was posed this way: *Several co-workers of yours have recently had personal belongings stolen from your workplace, and the company is planning to install high-resolution security cameras that use facial recognition technology to help identify the thieves and make the workplace more secure. The footage would stay on file as long as the company wishes to retain it, and could be used to track various measures of employee attendance and performance. Would this be acceptable to you, or not?* Some 54 % said the installation of surveillance cameras would be acceptable under these conditions; 24 % said it would not be acceptable; and 21 % said their views would depend on more details and context for the scenario.

A second scenario drew a very different response: *Your insurance company is offering a discount to you if you agree to place a device in your car that allows monitoring of your driving speed and location. After the company collects data about your driving habits, it may offer you further discounts to reward you for safe driving. Would that scenario be acceptable to you or not?* In this case, only 37 % said the bargain—my driving information in return for possible discounts—was acceptable to them; while 45 % said it was not acceptable; and 16 % said their agreeing to the deal would depend on their learning more details.

In each case, something of potential value was being offered respondents in return for the potential collection of personal information, but different people were comfortable with different deals. The conditions of the offer mattered to them and they weighed the value proposition differently, depending on the circumstances.

Part of the bargain people weigh when they are deciding if they like an information deal or not is how they feel about the party on the other side of the deal.

In the hierarchy of privacy concerns, Americans are not anxious to be known to or surveilled by hackers (the black-hat kind) or advertisers (Pew Research 2 2013). The next most sensitive area of sensitivity for people involves social surveillance. It is a more top-of-mind concern to people than government surveillance. People are more likely to experience or witness reputational privacy breaches within their own networks than they are to be aware of how the government's access to their data might negatively impact their lives (Chart 2).

One last example of how context colors Americans' views on privacy involves government surveillance programs themselves (Pew Research 5 2015). Far from being opposed to surveillance, the public generally believes it is acceptable for the

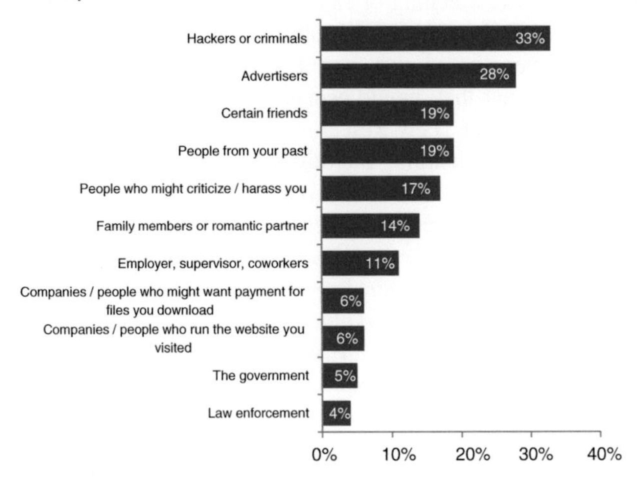

Chart 2 Who users try to avoid. *Source* Pew Research Center survey July 11–14, 2013

government to monitor many others, including foreign citizens, foreign leaders, and American leaders:

- 82 % say it is acceptable to monitor communications of suspected terrorists.
- 60 % believe it is acceptable to monitor the communications of American leaders.
- 60 % think it is okay to monitor the communications of foreign leaders.
- 54 % say it is acceptable to monitor communications from foreign citizens.

Yet, 57 % say it is *unacceptable* for the government to monitor the communications of U.S. citizens. At the same time, majorities support monitoring of those particular individuals who use words like "explosives" and "automatic weapons" in their search engine queries (65 % say that) and those who visit anti-American websites (67 % say that).

Implications for big data: Americans' views on these issues suggest there are ways for analysts of big data to make the case for their work. People are not instinctively opposed to data collection and use. They want to see what the tradeoff is, and under the right circumstances will accept the bargain. This might put some burden on big data analysts to make the case for their work and the benefits that will emerge from it, but it suggests that many are open to sharing information and being tracked if they understand what the upside of research is.

3. **Personal control and agency matter a lot to people.**

If the traditional American view of privacy is the "right to be left alone," the 21st Century refinement of that idea is the right to control their identity and information.

Computational Social Sciences

They understand that modern life won't allow them to be "left alone" and untracked, but they do want to have a say in how their personal information is used. There are several pieces of evidence for this. First, 86 % of internet users have tried to be anonymous online at least occasionally and 55 % of internet users have taken steps to avoid observation by specific people, organizations, or the government (Pew Research 2 2013).

At the attitudinal level, Americans say that being in control of who gets information about them. Even though they acknowledge that the boundary line between private and public information has sharply shifted toward "publicness" as the default condition of the modern moment, Americans continue to insist that they care about what happens to their personal information once it has been collected: 74 % say it is "very important" to them that they be *in control of who can get* information about them and 65 % say it is "very important" to them *to control what information is collected about them* (Pew Research 6 2015) (Chart 3).

Implications for big data: This basic American attitude about privacy is difficult to apply to big data. In many cases, the data are collected in ways that are not easy for users to control. All the greater burden falls, then, on researchers to show what they are doing and how they arrive at the insights they do. Perhaps applications of new online trust-building systems might help civilians assess and maybe contribute to the findings—that would including things like Reddit-style up- or down-voting schemes or allowing participant comments on the findings. That might give them a sense of agency and stake in the insights that the data generate.

4. **The young are more focused on networked privacy than their elders.**

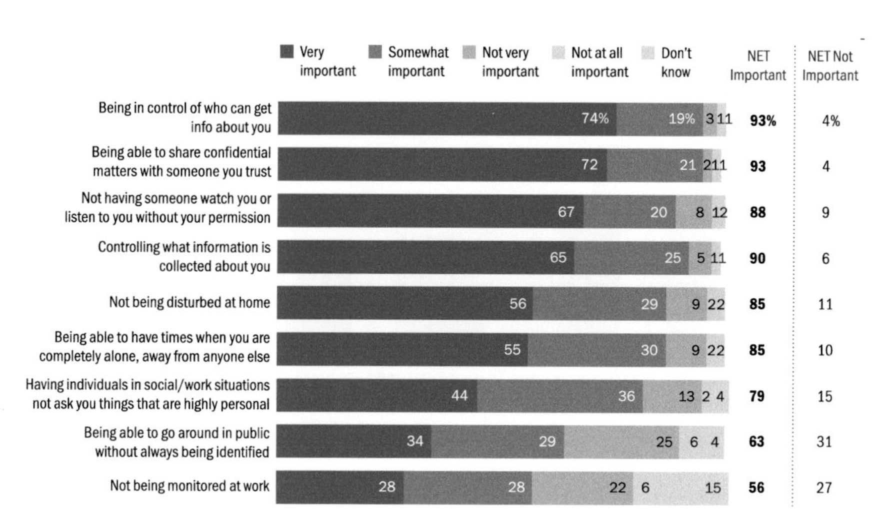

Chart 3 Americans hold strong views about privacy in everyday life. *Source* Pew Research Center survey January 27–February 16, 2015

Throughout the Pew Research data, those ages 18–29 are more likely than older adults to say they have paid attention to privacy issues, to have taken steps to protect their privacy, and to have suffered some kind of harm because of privacy problems (Pew Research 2 2013):

- They take steps to limit the amount of personal information available about them online—44 % of young adult internet users say this.
- They change privacy settings—71 % of social networking users ages 18–29 have changed privacy settings on their profile to limit what they share with others online.
- They delete unwanted comments—47 % social networking users ages 18–29 have deleted comments that others have made on their profile.
- They remove their name from photos—41 % of social networking users ages 18–29 say they have removed their name from photos that were tagged to identify them.

It is likely that this extra attention to personal reputation emerges from the reality that younger adults are more likely to know that personal information about them is available on line and to have experienced privacy problems. For instance, those 18–29 are more likely than older adults to have had an email or social media account hijacked or had online difficulties place them in physical danger (Chart 4).

The larger point here is that different people have different generational, cultural, or social circumstances which inform their attitudes and behaviors around disclosure or privacy protection.

Implications for big data: There is not a one-size-fits-all set of policies and solutions about how to handle big data. Researchers would help themselves by

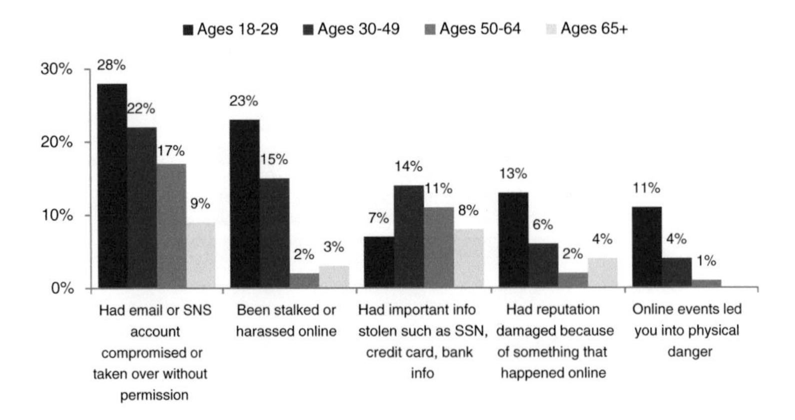

Chart 4 Young adults are the most likely to have had several—but not all—major problems with personal information and identity. *Source* Pew Research Center survey July 11–14, 2013

caring about different demographic, cultural, and generational sensitivities on privacy issues. One way to address this would be outreach to special segments of the population with assurances that the issues that matter most to them are on researchers' minds as they do their analysis and render their findings.

5. **Many know they do not know what is going on when it comes to the nature and scope of data collected about them.**

When it comes to their own role in managing the personal information, most adults are not sure what information is being collected and how it is being used. "I wouldn't know where to begin if I ever wanted to get to the bottom of what kind of profiles exist on me," one middle-aged suburban woman said in a Pew Research online focus group (Pew Research 7 2015). A 63-year-old man added: "Every organization wants at least pieces of me—what I buy, who I vote for, what movies I go to, what music is on my playlist, the medicines I take, how much energy I use, even who my friends are. It is impossible to imagine one part of my life that is not being documented by one company or another."

At the same time, many express a desire to take additional steps to protect their data online. When asked if they feel as though their own efforts to protect the privacy of their personal information online are sufficient, 61 % say they feel they "would like to do more," while 37 % say they "already do enough" (Pew Research 3 2014). When they want to have anonymity online, few feel that is easy to achieve. Just 24 % of adults "agree" or "strongly agree" with the statement: "It is easy for me to be anonymous when I am online."

Implications for big data: These findings underscore how fragile the relationship between big data analysts and the public are. People do not like surprises and will likely be unhappy if their data were used in ways they did not anticipate or that seem "out of the blue."

6. **Americans believe changes in law could make a difference, though their exact policy preferences are not fully clear.**

In the midst of all this uncertainty and angst about privacy, Americans are generally in favor of additional legal protections against abuses of their data. Some 68 % of internet users believe current laws are not good enough in protecting people's privacy online (Pew Research 2 2013); and 64 % believe the government should do more to regulate advertisers, compared with 34 % who think the government should not get more involved (Pew Research 3 2014).

When asked to think about the data the government collects as part of anti-terrorism efforts, 65 % of Americans say there are not adequate limits on "what telephone and internet data the government can collect."[2] Just 31 % say they believe that there are adequate limits on the kinds of data gathered for these

[2]Due to differences in the method of survey administration and questionnaire context, these findings are not directly comparable to previous Pew Research telephone surveys that have included a version of this question.

programs. The majority view that there are not sufficient limits on what data the government gathers is consistent across all demographic groups. Those who are more aware of the government surveillance efforts are considerably more likely to believe there are not adequate safeguards in place.

Relatedly, there is a striking divide among citizens over whether the courts are doing a good job balancing the needs of law enforcement and intelligence agencies with citizens' right to privacy: 48 % say courts and judges are balancing those interests, while 49 % say the courts are not (Pew Research 5 2015).

Implications for big data: It is easy to imagine that analysts of big data would gain public approval by helping people understand what is going on in the world of hyper-data collection and providing strategies and tools to help Americans regain a sense they are more knowledgeable about this environment and more competent to navigate it.

7. **Conclusion: The future of privacy**

When the Pew Research Center canvassed hundreds of technology experts and pundits about the fate of privacy in the coming decade, there were several themes in their predictions about the future that are relevant to the long-term viability of big data research (Pew Research 4 2014): First, Few individuals will have the energy or resources to protect themselves from "dataveillance." Privacy protection will likely become a luxury good. There will be technology tools and marketplace solutions that will be embraced by higher socioeconomic groups, but the capacity of average citizens to achieve privacy will diminish. Second, the prospect of achieving by-gone notions of privacy will become more remote as the Internet of Things arises and people's homes, workplaces, and the objects around them will "tattle" on them. Third, living a public life will be the new default. People will get used to this, adjust their norms, and accept more sharing and collection of data as a part of life—especially Millennials and the young people who follow them. Problems will persist and some will complain but most will not object or muster the energy to push back against this new reality in their lives.

In a way this is good news for the expansion of big data initiatives and those who would use them. Still, those who take advantage of these new realities bear the risk of pushing things too far and engendering a backlash if they do not accommodate insights from Americans' long history of asserting that privacy matters, there are ways that it is directly connected to social trust which is the bonding agent for any society, and there are parts of life that are best left unmonitored and protected from prying eyes.

References

Boyd, D. (2014) *It's Complicated.* Yale University Press. Downloaded on November 7, 2015 at http://www.danah.org/books/ItsComplicated.pdf

Pew Research 1. (2000, August 20). Trust and privacy online. Retrieved on November 2, 2015, from http://www.pewinternet.org/2000/08/20/trust-and-privacy-online/

Pew Research 2. (2013, September 5). Anonymity, privacy, and security online. Retrieved on November 7, 2015, from http://www.pewinternet.org/2013/09/05/anonymity-privacy-and-security-online/

Pew Research 3. (2014, November 12). Public perceptions of privacy and security in the Post-Snowden Era. Retrieved on November 7, 2015, from http://www.pewinternet.org/2014/11/12/public-privacy-perceptions/

Pew Research 4. (2014, December 18). The future of privacy. Retrieved on November 7, 2015 at http://www.pewinternet.org/2014/12/18/future-of-privacy/

Pew Research 5. (2015, March 16). Americans' privacy strategies Post-Snowden. Retrieved on November 7, 2015 at http://www.pewinternet.org/2015/03/16/americans-privacy-strategies-post-snowden/

Pew Research 6. (2015, May 20). Americans' attitudes about privacy, security, and surveillance. Retrieved on November 7, 2015 at http://www.pewinternet.org/2015/05/20/americans-attitudes-about-privacy-security-and-surveillance/

Pew Research 7. (2015, December). Privacy and information sharing (forthcoming).

Engaging the Public in Ethical Reasoning About Big Data

Justin Anthony Knapp

Ethical concerns in Big Data are of particular interest to a variety of professionals, researchers, and specialists. As the volume and variety of such data continue to grow rapidly, individuals from fields as different as geography and biomedical sciences to software engineering and sociology will have a wealth of new material available to further their academic and financial interests. All of these fields will have to adapt to the unique ethical issues related to digital privacy rights and the management of previously inconceivable amounts of data.

But researchers are not the only stakeholders in these issues. Certainly, academics and other professionals will have to devise new guidelines for their internal use and public policy may have to change rapidly but these modifications will be of limited value if the public at large does not understand or engage with the broader community of those who are gathering and using such data. Both to maintain the integrity of such data sets and to protect possibly vulnerable individuals, it is imperative that the ethical reasoning behind Big Data decision-making is a transparent and intelligible process. This chapter will attempt to discuss what some of the ethical issues are in Big Data collection and a theory of how to think about privacy rights. Throughout, examples will be given from both academic literature and everyday life which will hopefully encourage researchers to think about how they can communicate with the public about Big Data. A theoretical approach is adopted and applied to several intersecting segments of society with an emphasis on civic actors. Finally, a few practical suggestions will be given that may prove useful for ensuring the integrity of the data themselves as well as providing confidence to the public who are giving the data.

Theoretical frameworks for understanding violations of privacy and consent can come from civil liberties and social justice movements. One early approach can be taken from American lawyer Lawrence Lessig, whose 1999 book *Code and Other*

J.A. Knapp (✉)
Free culture and digital liberties advocate, Indianapolis, USA
e-mail: justinkoavf@gmail.com

J. Collmann and S.A. Matei (eds.), *Ethical Reasoning in Big Data*,
Computational Social Sciences, DOI 10.1007/978-3-319-28422-4_4

Laws of Cyberspace was written for a lay audience and published at a time when there was little existing legislation or public discussion on digital civil liberties. In that work, he suggests that there are four main regulators of online activity: Code (or Architecture), Laws, Markets, and Norms. There are inherent technical features of what technology we have and how they function which can generally be referred to as "Code". In the case of Big Data, that infrastructure has exploded into realms which were science fiction in 1999. For instance, security research firm Trend Micro created "GasPot" which is a digital honeypot used to simulate a gas pump (in computing honeypots are deliberately unsafe traps designed to attract malicious agents such as identity thieves or spammers). They found that this digital gas pump had 23 attacks on it in the course of a few months in early 2015 simply by virtue of being connected to the Internet. The clear implication is that actual gas pumps which accept our credit card information and have connections to security cameras are possibly just as vulnerable and definitely far more dangerous if compromised.

Big Data are collected, analyzed, and sometimes disseminated by private and well as public actors. For instance, the accumulation of market research data through social media has made it possible for ad companies such as Facebook and Google to create vast digital empires using business models that would have been impossible a decade ago. Additionally, government agencies collect huge quantities of data through telecommunications for the purposes of law enforcement, surveillance, epidemiology, and a host of other concerns. The technological tools that have allowed these organizations to gather and analyze these data are simply too sophisticated and change too rapidly for the public at large to give informed consent about the collection and use of such information. This requires the public to give a greater level of trust to these institutions than ever before, including many instances of implicit approval or simple blind faith in the best intentions of corporate and governmental organizations.

To use a simple example, virtually no users read software end user license agreements (EULAs) (Bakos et al. 2014). This problem has been apparent for over a decade and has resulted in serious breaches of privacy on the part of users who have made unwarranted assumptions about software providers being well-intentioned and who have also been intimidated by increasingly dense legalese used for increasingly long agreements. As this software becomes more deeply embedded in everyday life, it is unreasonable to expect that the average user will have a true understanding of what kind of information is being collected about him and how it will be used. The terms of such EULAs almost invariably favor sellers over users and buyers (Marotta-Wurgler 2011).

In the United States, many researchers could be considered members of the civic sector—they do not have the same priorities as private business interests nor public state actors, although at times have features of both or work with either. Civic sector organizations such as non-profits and co-operatives generally have greater degrees of trust placed in them in part due to the legal and practical demands of transparency that are placed upon them. Problems of trust are particularly acute amongst voluntary associations by their very nature as they cannot compel compliance like a state actor and they typically cannot provide the compensation of businesses. Since trust is a key element in the efficacy of institutions to perform (Newton and Norris 1999), it is necessary that researchers engender that trust by actively engaging the public when it comes to privacy concerns about Big Data; otherwise, they risk losing out on possible sources of data through non-compliance and disinterest.

Public perceptions of Big Data, privacy, and digital civil liberties in the United States can be dated to pre- and post-Edward Snowden whisteblowing. Reports vary about the ways in which Americans have modified their online habits as well as the extent to which they are concerned about issues related to pervasive surveillance. For instance, Preibusch (2015) concludes that although, "media coverage of [American domestic spying program] PRISM and surveillance was elevated for the 30 weeks following PRISM day, many privacy behaviors faded quickly". Alternatively, Schneier (2014) point out that over 700 million Internet users worldwide have taken some steps to change their behavior to avoid National Security Agency surveillance. The survey data conclude that in the United States and abroad, average Internet consumers definitely have concerns about personal privacy in Big Data from governmental (CIGI-Ipsos 2014; Rainie and Madden 2015) as well as corporate sources (Fisher and Timberg 2013). One possible explanation for this seeming disjunct between attitudes and behaviors is how difficult it is for typical Internet users to understand complex tools for protecting privacy. Since the polling data and common sense tell us that online services as well as portable and wearable computers are such ordinary devices for millions, they are unwilling or unable to forgo their convenience and ubiquity in spite of genuine, rational concerns about Big Data collection and retention.

In the face of this, some providers of online services have offered or mandated tools which are intended to make Internet users more secure. This is one part of reaching out to the public about ethical concerns in Big Data: using "push" technologies that force users to be more privacy-conscious. For instance, in 2015, the Wikimedia Foundation (WMF)—operators of several online educational resources, including Wikipedia—filed suit against the NSA and Department of Justice with representation by the American Civil Liberties Union. Their argument was that the hundreds of millions of users of the encyclopedia were harmed by indiscriminate collection of upstream data by the United States federal government. The case was dismissed due to lack of standing with the court arguing that the plaintiffs could not prove they were subject to upstream surveillance. How in principle they could prove this when the surveillance program is secret was not explained by the court. They appealed in 2016 and comparable filings by non-profits such as *Clapper v. Amnesty International US* have been dismissed on similar grounds.

The WMF also implemented HTTP Strict Transport Security (HSTS) on all Web traffic starting later that year, which is the culmination of a process begun years prior to Snowden's whistleblowing activities. This technology requires users to access to the site using an encrypted connection rather than plain HTTP which allows intermediary agents like Internet service providers (ISPs) to view traffic in plain text. Previously, users were simply given the option of viewing the encyclopedia with HSTS and in 2013, logged in users were required to use it as a part of editing. The average reader would not notice any difference aside from a small icon changing in a web browser but this policy seriously decreases the possibility of "man in the middle" attacks which allow faking of credentials. These attacks had become increasingly common across the Web in the first decade of the encyclopedia's existence.

As a 501(c)(3) charitable organization, they regularly publish transparency reports which inform users of federal government demands for data. They also helped to generate significant political action in 2012 by blacking out Wikipedia to protest SOPA and PIPA—proposed Congressional legislation that would have had a chilling effect on online communication and which would have imposed mandated snooping of users by ISPs. Subsequent attempts by law enforcement to deputize ISPs have been introduced regularly and have been defeated by a combination of political will and public outcry.

These attempts to engage the public on digital privacy are not limited to non-profits. One such example of a large Internet community operated as a for-profit is link-sharing message board Reddit, which is run as an independent company with former direct owner Advance Publications as a large shareholder. The site also participated in SOPA/PIPA blackouts and began publishing transparency reports in 2015, explicitly acknowledging privacy concerns related to online surveillance. They have also taken a hardline stance on harassing behavior including doxxing—the intentional leaking of personally identifying information about users. These details are published in order to shame or threaten others into silencing them from discussing controversial topics. Sometimes, this is done purely as prankery and other times it is to gain privilege to someone's real life, such as sending threats in the mail. The tension between free speech to share information and the concerns of vulnerable individuals and groups which require anonymity has played out in controversial message boards which site administrators deemed to be excuses for harassment and trolling. The site has banned user accounts and boards which existed solely to mock minority groups or which attempted to spread nude celebrity photos after high-profile data breaches. Similar struggles between free expression which culminates in abuse is not purely theoretical or confined to cyberspace—protests on college campuses about safe spaces and allegedly bigoted policies have been increasing in the United States for years. This is another valuable method of engaging the public: creating community norms and rules which demand respecting others' privacy. As it becomes understood as a part of simple etiquette that others have a reasonable expectation of privacy, the ethical principle behind that rule can be more easily enforced and encouraged.

A unique hybrid of a state agency cooperating with a non-state actor is the TOR Project. This organization was created in 2006 to manage the TOR Browser as well as other online communication projects which are based on "onion routing" initially developed by the United States Naval Research Laboratory. The history and technical specifications of the technology can be complicated but simply put, onion routing passes Internet communications through several encrypted layers—hence the metaphor of an onion. Since the data are bounced around many agents before reaching their final destination, it is extremely difficult to determine where a request originated and the only way to reliably de-anonymize this traffic is if a user does it himself, such as by accidentally associating his personal e-mail address with activities performed on the TOR software network. The initial purpose of the technology was for military intelligence but it has since grown into a vast network which runs a kind of parallel Internet sometimes known as the Dark Web which includes sites that can only be accessed via the TOR Browser. The Dark Web includes the same type of mundane information that anyone would anticipate on the Internet such as chat and search services but also allows for new opportunities for illicit drug sales, gun smuggling, fraud, and other illegal or gray market activities. It also allows for very secure communication between whisteblowers and news agencies, especially when it is paired with SecureDrop which lets users share files with one another.

The ironic twist is that law enforcement not only monitors this traffic heavily and conducts sting operations using it but actually provides much of the architecture that allows for this communication in the first place with computers that connect to the TOR network know as "exit nodes". Intelligence agencies are in the curious position of both making this highly secure communications technology and constantly trying to undermine its integrity. For instance, in 2014 a number of international law enforcement agencies cooperated to shut down Dark Web drug markets, particularly Silk Road 2.0 which usually operated by trading designer drugs for Bitcoin. In the process, it is suspected that the FBI paid researchers at Carnegie Mellon University to try to break the anonymity of the TOR network. The relationship between the public and civic sectors here becomes as complicated as the technical aspects of the software itself but the take-away for researchers who have highly sensitive data is that TOR paired with SecureDrop is an excellent way to transfer information. Additionally, it is very user-friendly, unlike other secure systems such as PGP for e-mail. In the words of the NSA itself, TOR is "the King of high-secure, low-latency Internet anonymity" with "no contenders for the throne in waiting".

Such attempts to engage the public are also not limited to online service providers. Manufacturers of mobile devices have also promoted more secure communication—for instance, Apple has included end-to-end encryption in recent models of iPhones and had a high-profile dispute in 2016 with the FBI regarding attempts to decrypt a phone associated with a mass shooting. The other major smartphone vendor is Google, whose Android operating system comes in several varieties across many devices, so their approach to encrypted communications cannot be as uniform as Apple but they have also insisted to the public as well as law enforcement agencies that the company has no way of accessing the content of messages sent on Android devices (and these claims are easier to substantiate since Android is made of free software that anyone can modify or audit for security concerns). They have issued similar warnings to users of their Chrome web browser and other services they provide. This is in the interests of both the end user and the company, who can claim plausible deniability about being liable for the content of any messages sent using these devices and services. If they cannot in a technical sense snoop on users, then they cannot be mandated to do so legally. This frees up resources that the company can use to be profitable rather than be deputized for surveillance.

One large lacuna in this discussion is connected devices which are not personal computers per se. Wearable devices including fitness trackers, home furnishings such as thermostats, and even the increasingly sophisticated computers in automobiles are all connected to Internet cloud service providers through the Internet of Things (IoT). It is possible that this vast array of mundane objects will provide far greater and more intimate information about users than even personal computers and smartphones. They also have unique vulnerabilities: researchers have shown through controlled tests that these devices can be hijacked remotely by third parties to cause motorists to lose control of their vehicles and critically important medical devices can be caused to malfunction. Fear of malicious hackers (i.e. "crackers") who want to take control of these systems may cause members of the public to be skeptical of helpful and professionally gathered data. The ironic side effect is that users will have given up masses of information involuntarily in their day-to-day lives but will not be willing to trust researchers who have good intentions for gathering data and professional standards for maintaining confidentiality. This can literally start from birth with connected devices such as baby monitors. Gao (2015) has shown that Americans already believe overwhelmingly that they want to control their personal information but cannot. This problem could be particularly acute as responsiveness to polling has been falling in America for several years prior to concerns about snooping. (Christian et al. 2012) Researchers who use wearable computers are encouraged to consider whether or not real-time data collection is necessary or appropriate for their projects. Some of these devices can collect and store data on the device itself and that data can be retrieved once the gadget is returned rather than broadcasting it.

These problems are apparent even to agencies which traffic in Big Data—in 2012, the NSA internally promoted a staffer to be the "Socrates of the National Security Agency" and write a column on ethical issues for other employees. One serious question already raised is that of consent. It is taken for granted in professional fields that handle private data that consent must be given to acquire such data and that a subject must be informed in order to give that consent, such as in Health Insurance Portability and Accountability Act (HIPAA) requirements for medical investigators and practitioners. But HIPAA requirements break down when paired with Big Data, as insurers and third parties can easily de-anonymize such data by comparing it to public records. And since such data are held by a variety of hospitals and health plans, they are vulnerable to attack by identity thieves who have a thriving market for pilfered medical records. (Chideya 2015) Vendors such as Microsoft are working on entirely new encryption and data management schemes specifically for the medical industry.

Returning to Lessig's framework, legal challenges to protecting digital civil liberties have become increasingly necessary in the United States. The executive and legislative branches either struggle to keep up with a changing landscape for digital privacy or are simply too invested in mining such data to want to constrain themselves from getting it—consider that the CIA even spied on Congress under provisions of the Patriot Act. Founded in 1990, the Electronic Frontier Foundation (EFF; which has included Lessig on its board) is one of the oldest digital rights organizations and has participated in several legal challenges to surveillance and other breaches of digital privacy. Although U.S.-based, they work globally to challenge repressive laws and partner with other such non-profits. There has perhaps been no other organization as successful in reaching the public regarding digital rights issues.

Lawsuits do not only target government agencies and outreach to the public must also include market-based solutions. The commercial world is deeply invested in Big Data and this creates an inherent tension between personal privacy and efficiency in firms. This dichotomy has been challenged by Calo (2015) who argues that markets actually rely upon privacy to function and that the Federal Trade Commission "has emerged as the de facto privacy authority in the United States". There need not be a disjunct between privacy and profitability in Big Data, especially since innovative products and services are created in market situations in order to *preserve* the integrity of such data. It is important to balance critiques of government surveillance with market reform in order to make a robust and durable culture of digital rights. (Paterson 2014) The division between legal- and market-based solutions also breaks down when we consider the great extent to which private and public entities cooperate on collecting and trading Big Data. One such collaboration is police departments in dozens of states which pay for the rights to databases generated by repo firms using license plate scanners. These devices take millions of pictures of license plates in parking lots and compare them to records that the company has to determine where a delinquent customer is in order to tow an automobile. These data are collected with virtually no effort and can be used to track the movements of almost anyone with an automobile.

But cultures do not exist based solely on tools, legal codes, and commerce. The final piece of Lessig's framework is norms—which by their very nature are not codified. As Lessig puts it in *Code*, "we live life subject to these norms… [they] constrain us in ways that are so familiar as to be all but invisible". These tacit feelings regarding privacy are sometimes the strongest ways of reaching the public regarding their digital rights. For instance, once a member of the public is shown how trivial it can be to de-anonymize Big Data or to make inferences based on what has been collected, this can cause a sense of having been violated. These inchoate expressions of outrage have been manifested in public demonstrations like Restore the Fourth rallies which were held throughout the United States roughly overlapping the Occupy movement but also come up in mundane situations such as when one swipes to view the next picture on someone's smartphone without permission. In order to effectively reach the public, it is necessary to capitalize on these justified perceptions which take abstract and sometimes overly complicated arguments about end-to-end encryption or monitoring of air-gapped laptops through heat signatures and makes these ethical concerns salient. Harnessing such feelings and making them persist into something more substantial is imperative.

What does this all mean for the research community? Academics do not have the power or resources of multinational corporations or governments, so the concerns raised above may seem irrelevant. For that matter, institutional review boards and professional standards act as watchdogs against malfeasance—they are not perfect instruments but they generally work well and create an ethical culture amongst academia. Big Data problems can still exist through misperception and a simple lack of understanding of best practices. If the horizon of the digital landscape moves too quickly for the public, then it certainly can for researchers as well who are deeply invested in their work and the systems that they have in place. Well-intentioned and competent investigators still have to be able to communicate to the public as well as their peers that Big Data are secure and that best practices are followed.

This chapter has largely focused on the United States but there is a serious tension between Americans and Europeans characterized by Kerry (2014) as, "conventional wisdom in Europe that Americans do not care about privacy". This can cause a serious rift and have legal implications. Take the example of Boston College and "The Belfast Project"—an oral history on The Troubles in Northern Ireland which was created in 2001. Researchers interviewed former IRA members whose personal stories included details about illegal activity. The interviewers assured the subjects that their stories would remain confidential until they died but those tapes were requested by the Police Service of Northern Ireland leading to a legal battle that has lasted for years. It is further complicated by the fact that the College has distanced itself from the Project organizers and the History Department claimed to be largely ignorant of it even existing until the 2010 publication of the book *Voices from the Grave*. The precedent this sets legally and culturally can have serious implications for researchers but it is a microcosm of a larger issue of trust between American and European institutions.

To use the framework that Lessig created, researchers can discuss Big Data privacy with the public by referring to Code, Laws, Markets, and Norms. In terms of Code, researchers can make sure that they are using best practices by consulting EFF publications and using some of the digital tools that they make available at no cost. Another excellent resource for electronic privacy is the Electronic Privacy Information Center (EPIC), which is a Washington, D. C.-based research center. Similar digital rights organizations exist globally such as Bits of Freedom in The Netherlands, South Africa's Right2Know, and the United Kingdom's Open Rights Group. One radical solution may involve simply not using digital records in the first place or digitizing print records for the purposes of analysis and then destroying the digital copies but retaining the print ones. This measure may prove impractical for many researchers but even simply having the computers which store Big Data be disconnected from the Internet is enough to make these records far more secure. Alternately, researchers can store large datasets on optical media or thumb drives. This division between devices which are connected and those which are not is not only a practical concern but one that helps to ease the fears of members of the public—as Hogan and Shepherd (2015) found, "control of the physical location of data centers shapes the possibilities of data agency and ownership". If you can display to the public how Big Data are stored in a different place and disconnected from the Internet, it can increase feelings of security in addition to the actual level of security itself.

In terms of Law, compliance is actually a double-edged sword. Since one of the main reasons for apprehension about Big Data collection is government snooping, it is important to communicate to the public that you have a transparency policy and explain what you do with data and with government requests for data. Researchers may wish to publish a warrant canary and similar statements about data integrity aimed at informing the public about the seriousness with which the academic community takes data sensitivity.

Regarding Market-based changes, researchers are encouraged to look for alternatives to common technologies which provide greater security and lower cost. Using computers which have free operating systems can allow for more flexibility and control over the settings of the device—examples include BSD and GNU/Linux rather than Mac OS X and Windows. This has the added bonus of encouraging further use of safer computer systems. The more common it is to use these operating systems, the more tools will be developed for them. Additionally, these software communities have volunteers who work on making fixes and taking suggestions on how to improve their work, so if there is a feature that you would like to see, you can suggest it. (Note that the free database program PostgreSQL is slated to include Big Data functionality.) Not all researchers will be able to invest the time in learning new operating systems nor will they always be able to control which devices they can use but it is worth discussing which options are legitimate with someone in an IT department who can likely offer some alternatives.

Finally, normative outreach to the public is powerful and can be accomplished through subtle means. The attitudes that Big Data researchers have are as important as any tools or laws. As Boyd (2010) has suggested, "the biggest methodological danger zone presented by our collective obsession with Big Data: *Just because data is accessible doesn't mean that using it is ethical*". [emphasis in the original] Having a friendly and approachable manner when discussing privacy engenders greater trust. There is a human element to Big Data that can sometimes be overlooked by an obsessive focus on computers and smartphones as well as assumptions that more data are always better (recall the problems of finding needles in haystacks).

Conducting research today is an exciting and dangerous prospect. Big Data exists in virtually every form about almost all of us and it is used in ways that were unimaginable in the past. It is paramount that the community of Big Data users and collectors remember that such data are not just "out there" somewhere but are the intimate details of real persons' lives and they have just as deep an interest in protecting their privacy as they do in the good work that is conducted with such data. It remains to be seen if the world has the wisdom and forbearance to follow the advice of Wau Holland: protect private data, use public data.

References

Bakos, Y., Marotta-Wurgler, F., & Trossen, D. (2014). Does anyone read the fine print? Consumer attention to standard-form contracts. *The Journal of Legal Studies, 43*(1), 1–35. doi:10.1086/674424.

Boyd, D. (2010). Privacy and Publicity in the Context of Big Data. Retrieved from http://www.danah.org/papers/talks/2010/WWW2010.html.

Calo, R. (2015). Privacy and Markets: A Love Story. Retrieved from http://papers.ssrn.com/sol3/papers.cfm?abstract_id=2640607.

Chideya, F. (2015). Medical Privacy Under Threat in the Age of Big Data. Retrieved from https://firstlook.org/theintercept/2015/08/06/how-medical-privacy-laws-leave-patient-data-exposed/.

Christian, L., Dimock, M., Doherty, C., Keeter, S., & Kohut, A. (2012). Assessing the Representativeness of Public Opinion Surveys. Retrieved from http://www.people-press.org/files/legacy-pdf/Assessing%20the%20Representativeness%20of%20Public%20Opinion%20Surveys.pdf.

CIGI-Ipsos. (2014). CIGI-Ipsos Global Survey on Internet Security and Trust. Retrieved from https://www.cigionline.org/internet-surveyl.

Fisher, M. & Timberg, C. (2013). Americans Uneasy About Surveillance But Often Use Snooping Tools, *Post* Poll Finds. Retrieved from https://www.washingtonpost.com/world/national-security/americans-uneasy-about-surveillance-but-often-use-snooping-tools-post-poll-finds/2013/12/21/ca15e990–67f9-11e3-ae56-22de072140a2_story.html.

Gao, G. (2015). What Americans Think About NSA Surveillance, National Security and Privacy. Retrieved from http://www.pewresearch.org/fact-tank/2015/05/29/what-americans-think-about-nsa-surveillance-national-security-and-privacy/.

Hogan, M., & Shepherd, T. (2015). Information ownership and materiality in an age of big data surveillance. *Journal of Information Policy, 5*, 6–31. doi:10.5325/jinfopoli.5.2015.0006.

Kerry, C. (2014). Missed Connections: Talking with Europe About Data, Privacy, and Surveillance. Retrieved from http://www.brookings.edu/~/media/research/files/papers/2014/05/20-europe-privacy-surveillance-kerry/kerry_europefreetradeprivacy.pdf.

Marotta-Wurgler, F. (2011). Some realities of online contracting. *Supreme Court Economic Review, 19*(1), 11–23. doi:10.1086/664560.

Newton, K., & Norris, P. (1999). Confidence in Public Institutions: Faith, Culture, or Performance? Retrieved from http://www.hks.harvard.edu/fs/pnorris/Acrobat/NEWTON.PDF.

Paterson, N. (2014). End user privacy and policy-based networking. *Journal of Information Policy, 4*, 28–43. doi:10.5325/jinfopoli.4.2014.0028.

Preibusch, S. (2015). Privacy Behaviors After Snowden. *Communications of the ACM, 58*(5), 48–55. doi:10.1145/2663341.

Rainie, L., & Madden, M. (2015). Americans' Privacy Strategies Post-Snowden. Retrieved from http://www.pewinternet.org/files/2015/03/PI_AmericansPrivacyStrategies_0316151.pdf.

Schneier, B. (2014). Over 700 Million People Taking Steps to Avoid NSA Surveillance. Retrieved from http://www.lawfareblog.com/over-700-million-people-taking-steps-avoid-nsa-surveillance.

The Ethics of Large-Scale Genomic Research

Benjamin E. Berkman, Zachary E. Shapiro, Lisa Eckstein and Elizabeth R. Pike

1 Introduction

Over the past few years, there has been a dramatic evolution of our technological ability to gather and share information. This has enabled the collection, distribution, and analysis of vast amounts of data, in ways never before possible. While there has not been a distinct watershed moment, this type of increasingly large data collection has come to be known as "big data," and is defined by the National Science Foundation as involving "large, diverse, complex, longitudinal, and/or distributed data sets generated from instruments, sensors, Internet transactions, email, video, click streams, and/or all other digital sources available today and in the future." (National Science Foundation 2010).

While big data has applications in many fields, its potential in the biomedical research settings is among the most exciting. Through the creation of big data health repositories, researchers are able to gather information from a multitude of clinical and research sources, greatly expanding the breadth of their data, while allowing them to more widely share information with other researchers (Bollier and Firestone 2010). This enables researchers to study conditions in an entirely new way, ideally allowing advances to be made more quickly. Crucially, big data facilitates more

B.E. Berkman
Department of Bioethics, Clinical Center, and Bioethics Core, National Human Genome Research Institute, National Institutes of Health, Bethesda, Maryland, USA

Z.E. Shapiro
Harvard Law School, Cambridge, Massachusetts, USA

L. Eckstein
The Faculty of Law, University of Tasmania, Tasmania, Australia

E.R. Pike (✉)
Presidential Commission for the Study of Bioethical Issues, Washington, DC, USA
e-mail: elizabeth.pike@bioethics.gov

J. Collmann and S.A. Matei (eds.), *Ethical Reasoning in Big Data*, Computational Social Sciences, DOI 10.1007/978-3-319-28422-4_5

efficient analysis of data by allowing complex work to be spread out across multiple investigative sites. Additionally, by allowing data to be aggregated across many different investigational sites, big data allows researchers to solve challenging or rare health problems that had previously proven difficult to investigate.

This potential for big data to advance our understanding of human disease has been particularly heralded in the field of genomics. Recent technological advances have accelerated the massive data generation capabilities of genomic research. Next-generation sequencing techniques now use semiconductors and nanotechnology that increase the speed with which genomes are sequenced, resulting in a dramatic reduction in the time needed to sequence a given genome. This has allowed researchers to undertake larger scale genomic research, with significantly more participants, further spurring the generation of massive amounts of data. The advance of technology has also triggered a significant reduction in cost, allowing large-scale genomic research to be increasingly feasible, even for smaller research sites. This trajectory is likely to continue, as researchers predict that more advanced DNA sequencing technologies will be able not only to generate terabase-scale sequence data in seconds, but they will be able to sequence genomes for little or no cost (Schadt 2012). Along with more advanced methods of sequencing genomes, there have been improvements in the methods for collecting, storing, and sharing the data, particularly using computer-based databases, which have facilitated the rise of big data in genomics. We will use the term Large Scale Genomic Repositories (LSGRs) to refer to these research resources. The rise of genetic research has triggered the creation of many LSGRs, some of which contain the genomic information of more than a million research participants.

While LSGRs have genuine potential, they also have raised a number of ethical concerns. Most prominently, commentators have raised questions about the privacy implications of LSGRs, given that all genomic data is theoretically re-identifiable. Privacy can be further threatened by the possibility of aggregation of data sets, which can give rise to unexpected, and potentially sensitive, information. But beyond privacy concerns, LSGRs also raise questions about participant autonomy, public trust in research, and justice. In this chapter, we explore these ethical challenges, with the goal of elucidating which ones require closer scrutiny and perhaps policy action.

2 The Promise of LSGRs

While all scientific research produces data, genomic analysis is somewhat unique in that it inherently produces vast quantities of data. Every human genome contains roughly 20,000–25,000 genes, each comprised of over three million base pairs, so that even the most routine genomic sequencing or mapping will generate enormous amounts of data (International Human Genome Sequencing Consortium 2004). Since most studies include many different individuals, each with their own unique genomes, sequencing genomes of groups or populations produces huge quantities of data for researchers to analyze.

LSGRs are not merely a useful tool in organizing and compiling genetic research. Genomic research is a natural fit for big data, due to the complex nature of gene-based therapies and investigations, which necessitate the study and comparison of many individual genomes. For common diseases, it has become clear that a range of genomic variants can play a part in determining a given individual's risk for certain diseases or particular health outcomes. In order to find those variants, each of which might only make a small contribution to a given health risk, researchers must study a large number of both healthy and affected individuals, in order to identify the relevant genomic differences.

The vast quantity of data generated by such an analysis would once have overwhelmed even the most well-funded research labs. However, the use of LSGRs has enabled widespread data sharing, allowing analysis efforts to be spread across any number of investigational sites. This reduces analytic bottlenecks, while permitting more timely data analysis than any one investigative team would be able to accomplish on their own.

Aggregation also facilitates the study of rare diseases, where it is often difficult to find and recruit sufficient numbers of subjects with the relevant condition. LSGRs facilitate the collection of data from a geographically broad range of research sites, allowing advances in understanding that would be impossible to produce from studying small groups of individuals. By allowing data aggregation and pooling of data from many investigational sites, genetic underpinnings of various conditions can be identified, allowing researchers to begin the search for targeted therapies to combat some of the most devastating, and rare, genetic based conditions. LSGRs provide adequate statistical power to address questions that were previously infeasible due to logistical and funding limitations. Aggregation of disparate data sets also can allow researchers to make novel connections, or reveal trends not readily apparent in any one data set.

Given the potential of LSGRs to advance our understanding of disease, it is easy to understand why scholars predict that the use of LSGRs will only accelerate in the coming years. Indeed, there are already signs that LSGRs will become an increasingly common feature of the research landscape. In particular, a recent NIH genomic data-sharing policy requires that any researcher who receives funding for the production of genomic data must deposit their sequence data in a central repository (Genomic Data Sharing 2014). Policies like this are the first step in creating more widespread and informative LSGRs, and indicate that LSGRs may become a common feature of any significant genomic research.

Beyond the 2014 NIH genomic data-sharing policy, there are several examples of well-funded, emerging LSGRs that have already contributed significantly to our understanding of genomics and human disease. One example is the Million Veteran Program (MVP), started by the Department of Veterans Affairs in May of 2011. The MVP contains genomic, and some clinical information, from veterans who receive their care from VA *and* who volunteer to participate in the program. The initial benefits of this database are already being realized, with the VA using this information to identify patterns of illness following deployment.

Additionally, President Obama's "Precision Medicine Initiative" includes as a centerpiece a national repository containing health records and genomic sequence data from more than one million volunteers. The hope is that such a database will allow researchers to study the mechanisms by which peoples' genes, environment, and lifestyle affect their health, in ways not possible without the pooling of large amounts of data. By combining genomic information into population studies, hidden genomic influences may be identified. Beyond potentially revealing the causes of various conditions, this could elucidate opportunities for targeted therapies, allowing the development of cures with maximum efficacy.

3 Privacy and Re-identification

Despite LSGRs' promise for scientific advancement, their increasing ubiquity raises considerable privacy challenges (Lane et al. 2014). Most genomic samples and data are included in LSGRs premised on a promise of anonymity. A major concern is that this promise might be undermined by the possibility of re-identification (Rothstein 2010). While technically very difficult, re-identification can occur when researchers apply bio-informatic techniques that cross-reference existing, identified data sets with the genomic information contained in the LSGRs. These concerns are far from theoretical. Indeed, several groups of researchers have demonstrated that re-identification is possible, even with the limited information contained in de-identified LSGRs. In a seminal study led by Gymrek and colleagues, researchers were able to discover the identity of some individuals whose genomes had been sequenced as part of a genomics project. The research team wrote an algorithm that was able to infer an individual's array of genetic markers, called a haplotype, from the nucleotide sequence of his Y chromosome. The team then searched genealogical databases for the names of men with corresponding Y-chromosome haplotypes, and, after cross-referencing the last names with publicly available records, correctly identified several individuals (Gymrek et al. 2013).

Another study utilized public databases, which make genome-scale RNA abundance profiles (which reveal the amount of RNAs in different cells) available to anybody with the internet. Researchers were able to generate DNA barcodes from these data, which could be screened against DNA databases kept by government agencies (to identify DNA samples associated with unsolved crimes for example). It is possible that comparing these data sets could reveal the identity of a research participant. In 2012, Schadt and colleagues utilized RNA abundance measurements to infer a DNA-based barcode that was specific enough to re-identify individuals whose data was part of a collection of hundreds of millions of individual genotypic profiles obtained in a completely different research context (Schadt et al. 2012). Researchers have also reported that a personal large-scale SNP genotypic profile is sufficient to resolve whether an individual participated in a specific genome-wide association study, even if the study reports only summary statistics such as allelic frequencies (Homer et al. 2008).

With re-identification existing as an increasingly real possibility, attention has shifted to the challenges associated with offers of anonymity in genetic research. Re-identification concerns are heightened further by the aggregation of ever-greater amounts of information on the internet. This aggregation problem creates a novel threat to privacy, as cross-referencing this information with LSGRs can give rise to unexpected and potentially sensitive inferences and information. Furthermore, recent research has raised the possibility that scientists could use genetic markers from DNA in order to create a fairly accurate picture of an individual's face, highlighting that we are only beginning to realize some of the privacy implications raised by access to genetic information (Claes et al. 2014).

The above discussion highlights the potential for re-identification of genetic research participants. However, a more nuanced understanding of the *risks* that such re-identification poses to participants warrants closer scrutiny. A helpful way of assessing such risks is separating out participants' welfare and non-welfare interests (Tomlinson 2009). Welfare risks are best thought of as individual direct harms that represent a real personal risk to the individual. In contrast, non-welfare risks do not present a risk of immediate personal harm, but rather represent abstract harms to an individual's wishes, desires, or preferences. A non-welfare risk can be said to occur when an individual loses control over their personal information (Tomlinson 2009). We address these different kinds of harms separately in the next sections.

4 Welfare Interests

Genomic big data research may expose subjects to psychological, social, and economic harms, particularly if the research reveals sensitive information about re-identified individuals, or racial/ethnic/geographic groups with which they identify. Psychological harms include undesired changes in thought processes and emotion (e.g., episodes of depression, confusion, feelings of stress, guilt, and loss of self-esteem). Social and economic harms might include embarrassment within a participant's business or social group, loss of employment, or criminal prosecution caused, for example, by invasions of privacy and breaches of confidentiality. Additionally, some social and behavioral research may yield information about individuals that could "label" or "stigmatize" the subjects, either as individuals or through association with a specific group. While these harms are often cited as reasons to worry about genomic research, evidence of these harms is thus far quite low.

4.1 Psychological Harms

Arguments about psychological harms assume that research participants will be given distressing information about their genetic health risks, which will cause undue negative emotions. There is a robust psychological literature, however, that suggests that people are more emotionally adaptable than they think, and that we are terrible at affective forecasting, or predicting our future emotional reactions to negative events. While we often assume that learning about genetic risk for serious diseases will be devastating, in reality, the data suggest that the negative psychological effects of learning such information are generally transient and mild. This has been attributed to two psychological concepts: immune neglect and the focal illusion. Immune neglect refers to "the failure to anticipate how easily and quickly we make sense of and adapt to negative events." (Peters et al. 2014). The related focal illusion bias "is the tendency to focus on the affective consequences of a single, focal future event, while ignoring the emotional impact of non-focal events on well-being." (Peters et al. 2014).

The minimal psychological impact of negative genetic information has been demonstrated in a range of contexts (Heshka et al. 2008). For example, the REVEAL studies (Risk Evaluation and Education for Alzheimer's disease) were the first randomized controlled trials designed to evaluate the impact of susceptibility testing using the Alzheimer's Disease ("AD") susceptibility gene *APOE-ε4*. These comprised a series of four multi-site, randomized clinical trials examining psychosocial and behavioral responses to genetic risk assessment for AD using *APOE* disclosure (Roberts et al. 2011). The studies found little negative emotional impact (Green et al. 2009). Another systematic review similarly found no increased distress within the year after testing, and actually demonstrated a decrease in stress for many participants post-test (Broadstock et al. 2000). Similarly, a review of the literature on responses to genetic testing of cancer susceptibility found that there was very little evidence of adverse psychological effects observed among people who learn that they have a genetic predisposition to certain cancers (Meiser 2005). Similar data exists for testing range of other conditions, including Huntington's disease, breast cancer, and colon cancer, among others.

While we do not mean to minimize the possibility of psychological harms resulting from disclosure of genetic risk information, the existing literature should force us to consider whether our society is "systematically overestimate[ing] the durability and intensity of the affective impact of events on well-being." (Peters et al. 2014). Our argument is merely that policy makers and the scientific community should be cautious about using the psychological concerns of receiving genetic test results to justify regulations that will have a profound impact on the scientific enterprise.

4.2 Discrimination

Genetic discrimination ("GD") commonly refers to "the differential treatment of asymptomatic individuals or their relatives on the basis of their real or assumed genetic characteristics." (Otlowski et al. 2012). Differential treatment can occur within interpersonal and institutional domains, but institutional domains have been the focus of regulatory efforts. Objective evidence of GD has been difficult to establish and, until recently, its prevalence and depth has been largely undocumented.

Some studies have even presented positive evidence suggesting skepticism about GD's scope. A U.S. study on insurance outcomes published in 2009 surveyed 47 unaffected individuals with a genetic predisposition to breast cancer, concluding, "[r]esults suggest fear of GD is prevalent, yet data do not support evidence that GD exists." (McKinnon et al. 2009). Two adverse events were reported to have occurred when individuals changed health insurance. The study found no reports of job discrimination due to genetic status or family history of cancer. Furthermore, we are not aware of any instances in which GD has arisen from genetic research projects. In the closest available report, Kathy Hudson and others reported a case study in 1995 in which a research geneticist determined—outside the context of a research project—that a four-year old boy carried a genetic alteration that causes long QT-syndrome. His father subsequently was unable to obtain insurance coverage for his son because of this mutation (Hudson et al. 1995).

In the U.S., early experience with the Genetic Information Non-Discrimination Act (GINA) similarly suggests that perhaps there is less cause for concern than previously thought. Enacted in 2008, GINA was passed as a way to combat fears that genetic discrimination was a barrier to adoption of clinical genetic testing (Prince and Berkman 2012). The law works both prospectively (prohibiting employers and health insurance companies from receiving genetic information) and retrospectively (punishing bad actors who have illegally used genetic information as the basis for employment or actuarial decisions). While a watershed achievement, there have been remarkably few cases brought under the law (Genetic Information Non-Discrimination Act Charges 2014). Since 2010, there has been an annual average of just 48 cases reaching merit resolution and damages have not been substantial, averaging less than $1 million in total annual awards. While there have been more documented instances of discrimination in the life insurance and long-term care insurance areas, a systematic review of existing data led researchers to conclude that no policy intervention is currently justified, concluding that "with the notable exception of studies on Huntington's disease, none of the studies reviewed here (or their combination) brings irrefutable evidence of a systemic problem of GD that would yield a highly negative societal impact." (Joly et al. 2013).

As with the discussion of psychological harms, we do not mean to minimize the problem of genetic discrimination. It is certainly possible that genetic discrimination could eventually become a serious problem. While policy-makers should be cautious about imposing burdens on the research enterprise when there is little evidence of a current widespread problem, there is reason to guard against dismissing GD too quickly. As genetic information becomes more available and as our knowledge of the links between phenotype and genotype improves, insurance companies and others may take the opportunity to incorporate the information into decision-making. In one study, researchers at Georgetown University asked underwriters from insurance companies to underwrite hypothetical applicants who had received a genetic test result indicating increased risk of a future health condition. In seven of 92 total decisions, underwriters said they would deny coverage, place a surcharge on premiums, or limit covered benefits based on an applicant's genetic information. Adverse determinations were dispersed among the surveyed underwriters, across the hypothetical examples and despite relevant state-level proscriptions on genetic discrimination (Politz et al. 2007).

5 Non-welfare Interests

5.1 Trust

Even though there might not be current evidence of extensive individual welfare harms, one still must be concerned about the threat of harm to the non-welfare interests of participants. This can result from the lack of control over their samples and data, as well as the harm of broken promises, as participants participate in research with the expectation that they will not be personally identified by the data, and that their data will not be publically linked to them. Even if this does not result in tangible economic or mental harm to the individual, participant's non-welfare interests can be harmed by the release of this information. For these reasons, maintaining trust in the research enterprise and in the process of developing LSGRs is fundamental to the ongoing success of LSGRs and the research enterprise. And yet, the way that LSGRs are currently being created falls short of best practices for establishing and fostering trust.

Although some of the samples stored in LSGRs are collected from people who have provided consent for the genetic material to be used in a wide array of research projects, in other cases, the samples stored in LSGRs were collected without the source's knowledge or consent. For example, researchers often rely on samples collected as medical waste—blood or bodily tissue obtained in the clinic in excess of what was strictly necessary for testing or diagnosis. Current U.S. laws and guidelines allow the excess medical waste to be collected and stored in LSGRs without the source's knowledge or consent. Collection and storage of medical waste is generally governed by the Health Insurance Portability and Accountability Act's Privacy Rule, which places only limited restrictions on the ability to collect and store medical waste without consent. When samples are de-identified, the Privacy Rule places no restrictions on their use or disclosure. Even samples stored with specific identifiers can be used or disclosed under the Privacy Rule if the information is released as part of a "Limited Data Set" or if an Institutional Review Board has waived the requirement that individuals provide informed consent.

Researchers also rely on samples collected through the process of newborn screening—a public health screening process whereby newborns' heels are pricked and blood is collected and tested in the first few days of the child's life. Newborn blood spots are thought to be "an especially rich source of research material: they are stable over time, they constitute an unbiased collection of samples since they represent the entire population, and they can potentially be linked to basic demographic information" (Suter 2014). In many cases, the collection of a newborn's blood occurs without the parents' knowledge or consent (Suter 2014). The samples are then retained, in some cases indefinitely, for a range of subsequent uses (Citizens' Council on Health care 2009). The research use of newborn samples accelerated in 2009 due to an NIH grant that funded the *Newborn Screening Translational Research Network*, a national repository of newborn blood samples for use in research (Scutti 2014).

Although the federal Common Rule governing research with human subjects generally requires that investigators obtain informed consent from research participants, consent often is not required for research involving genetic samples. First, to the extent research samples are de-identified, the research is not considered human subjects research at all such that the Common Rule requirements (including the requirement of informed consent) do not apply. Second, even research using identifiable biospecimens may nevertheless be exempt from the Common Rule requirements of informed consent if data is "recorded by the investigator in such a manner that subjects cannot be identified, directly or through identifiers linked to the subjects." Third, even if the research is not considered exempt, an IRB is permitted to waive the requirements for informed consent in certain circumstances. Research using identifiable biospecimens can often qualify for waiver because the sheer number of people from whom genetic data has been collected renders re-contact and obtaining informed consent impracticable or impossible (Geetter 2011).

Given the lack of legal limitation, it is unsurprising that there are vast numbers of samples that are likely to have been collected without people's knowledge or consent. As of 1999, the RAND Corporation estimated that U.S. research repositories contained 307 million tissue samples. These samples were taken from 178 million individuals, accounting for almost two-thirds of the American population (Eiseman 2000). The RAND report conservatively assumed that the number of samples would grow by 20 million per year, which would mean that more than 600 million samples are being stored today, which does not even fully account for new sources of biological samples (direct-to-consumer genetic testing, criminal databases, etc.) that were just emerging at the end of the 20th century. It does not seem like much of an exaggeration to conclude, therefore, that "virtually everyone has his or her tissue on file" (Dunn 2012).

The potential for loss of trust in LSGRs when people learn that their genetic material has been collected, stored, and used without their knowledge or consent is high, and hugely consequential. This loss of trust has already occurred at the state level. In two states, Texas and Minnesota, parents learned that blood samples from their newborns had been collected without their consent and had been stored and used for a range of purposes including research. They subsequently brought suit. The Texas lawsuit, *Beleno v. Texas Department of Health Services*, ultimately led to the state agreeing to incinerate approximately 5.3 million newborn blood samples (Waldo 2010). The Minnesota lawsuit, *Bearder v. Minnesota*, ended with the state agreeing to "destroy all blood samples in long-term storage … and to pay nearly $1 million in legal costs." (Olson 2014).

These cases go beyond potential legal and financial consequences to highlight the less tangible ramifications of insufficiently informing and accommodating the views of potential participants in large-scale genetic research. Notably, Andrea Beleno, the named plaintiff in the Texas lawsuit, stated that she might have consented to the collection and subsequent use of her newborn's genetic data if she had trust in the enterprise: "If they had asked me … I probably would have consented. The fact that it was a secret program really made me so suspicious of the true motives, there's no way I would consent now" (Roser 2009). Surreptitious collection and use—collecting and using samples without the knowledge and consent of the source—leads to lack of trust in the enterprise. Without trust in the mission of LSGRs, LSGRs are at risk of the type of lawsuits that resulted in incineration of millions of samples along with a more widespread loss of faith in the medical research establishment more broadly.

5.2 Autonomy

Informed consent is a cornerstone of research ethics. However, LSGRs have forced a reexamination of existing regulations and norms. Traditionally, there was a clear distinction between data that included identifiers (e.g., name, date of birth, social security number, etc.) and data that had been de-identified. Under the Common Rule, secondary research involving de-identified data has not been considered to be human subjects research, and thus has not required IRB review. This regulatory distinction ultimately meant that consent has not been required for much of the genomic data contained in research repositories.

In large part because of concerns about re-identification of genomic data, proposed changes to the Common Rule look to obliterate the distinction between identified and de-identified data (Federal Policy for the Protection of Human Subjects 2015). The net effect of this change will be to require some kind of informed consent for any sample or data that will be used for research. While adopting a posture that seems more respectful of individual autonomy, this change could have a profound effect on the research enterprise generally, and on LSGRs in particular. The proposed rules would likely only apply prospectively, and would introduce the requirement of consent to collecting samples and data for subsequent use where one did not exist before.

Implementation of this new requirement will depend, in part, on whether participants are willing to accept the idea of blanket or broad consent. Blanket consent refers to the notion that a participant could give their consent at a single interaction, but would give permission for ongoing, open-ended use. Broad consent is similarly non-specific, but includes provisions wherein future uses are subject to some constraints (e.g., not for morally controversial topics, such as cloning). Blanket and broad consent can be compared to other approaches that require more study-specific consent, which obviously provide more information to a potential participant, but at significant cost to the research enterprise (Grady et al. 2015).

Some form of broad consent is expected to be part of the revisions to the Common Rule. Furthermore, there is evidence to suggest that participants are willing to accept such an approach. While a complete analysis is beyond the scope of this chapter, the data seems to indicate that individuals want to be asked for their permission once, but do not need to be approached to provide consent for specific subsequent uses (Wendler 2006; Chen et al. 2005). In fact, in one recent survey of various consent models for the use of stored genetic samples, potential participants viewed real-time specific consent as the least desirable option (Tomlinson et al.

2015). Unfettered blanket consent was also not widely supported, with subjects seeming to prefer the broad consent model where one-time permission is given, but when there are limits on controversial research uses, or a mechanism to withdraw at any point.

Any informed consent paradigm will involve some tradeoffs between burden on the research enterprise and participants' ability to exercise control over the use of their samples. As LSGRs proliferate, it seems untenable to continue with the status quo, where research is being conducted on samples and data without participant knowledge or consent. However, in the interest of minimizing burden on the research enterprise, careful consideration should be given to the rules that will be imposed. If implemented thoughtfully, broad consent seems like it could be an acceptable and appropriate compromise between respecting autonomy and facilitating research.

In addition to prospective consent, two additional autonomy-related concerns are raised by the proliferation of LSGRs. First, there are retrospective questions about the appropriateness of using genomic data and samples when there is inadequate or problematic evidence of consent. We term this the "grandfathering problem." When researchers seek to access genetic samples, many of which might be very old, how much evidence of high quality informed consent is required before allowing research to be conducted? For instance, perhaps a researcher retires and transfers a career's worth of samples to a biobank. Some of those samples might have been collected before modern informed consent laws and norms were in place, meaning that consent has not been documented, or is non-existent for *any* form of research with the samples. Or perhaps some of those samples were collected for a specific research purpose, and the consent form never mentioned the possibility of any sort of genetic research methodology (or mentioned only rudimentary forms of genetic analysis) suggesting that consent could be inadequate. Even more challengingly, some of those samples might have been collected from vulnerable populations (e.g., prisoners, psychiatric patients, adults lacking capacity, etc.).

Given that norms and rules evolve, we cannot simply apply today's consent standards to yesterday's samples and data. On the other hand, it seems ethically problematic to knowingly use research resources of questionable provenance. Important conceptual work will have to be done to develop an ethical framework that considers a number of relevant factors. First, we need to establish the extent to which inadequate or missing informed consent is ethically problematic in a range of scenarios. For example, having firm evidence that samples were collected from vulnerable individuals without consent raises more concerns than a mere lack of documentation of informed consent. Second, we need to decide how strongly to weigh the feasibility of obtaining additional, present-day consent for subsequent research use as a way of demonstrating respect for individual autonomy against the additional burdens placed on the research enterprise. It is appropriate to seek re-consent in certain situations, but there should be limits on the burdens imposed on the research enterprise. Finally, we need to explore the weight that we are willing to give to the unique qualities or irreplaceable scientific value that a given set of samples or data might possess.

We suggest that an appropriate balance between these three factors would allow questionable samples and data to be grandfathered only in cases where the unique scientific value outweighs the relevant ethical concerns. As one possible model, the National Human Genome Research Institute has instituted a policy stating that as of a specific date, previously collected samples can continue to be used for genomic data sharing as long as the existing consent forms are not inconsistent with such use. In order to discourage researchers from only using previously collected samples indefinitely, this rule only remains in place for five years. After that time, researchers will need a strong scientific justification to continue using samples that were not obtained with specific consent for broad data sharing.

The final autonomy-related concern exacerbated by the proliferation of LSGRs relates to the right to withdraw from research. Enrolling in research is not just a one-time decision; it is a well-established principle of research ethics that participants have the right to withdraw from participation at any time. In the context of actual physical participation in research, this is conceptually straight-forward as an individual can choose not to show up or to leave the study premises. But in the context of LSGRs, where data are being shared widely throughout the research community, withdrawal can be difficult or impossible. LSGRs should be designed such that individuals retain some ability to pull their information back should they choose. However, once the data has been widely shared, absolute eradication of data might not be feasible. LSGRs should prompt a re-examination of what the right to withdraw from research actually entails, and should encourage construction of consent forms that manage participant expectations accordingly.

5.3 *Justice*

There are two primary justice concerns arising out of LSGRs. The first relates to the unfortunate lack of diversity in genomic medicine. While genomic research has been presented as an important tool for unlocking the potential of genomic medicine, research efforts thus far have focused almost exclusively on people of European descent. For example, as of 2011, less than 10 % of participants included in genome-wide association studies ("GWAS") were not of European decent (Rotimi 2012). In the U.S., one study found that 92 % of GWAS participants were white, and only 3 % were African-American (Haga 2010). The worry is that without a broader racial and ethnic focus, researchers will develop a skewed understanding of which variants are relevant to human disease. Genotype-phenotype associations will be less generalizable for underrepresented populations, meaning that the majority of medical benefits will flow to an already advantaged segment of our global population. As Carlos Bustamante and colleagues stated:

> It is tempting to focus on populations that are motivated, organized, medically compliant and otherwise easy to study. But by failing to develop resources, methodologies and

incentives for underserved people, we risk perpetuating the health disparities that plague the medical system. Those most in need must not be the last to receive the benefits of genetic research. (Bustamante et al. 2011).

In order to avoid exacerbating health care inequality, LSGRs need to focus on engaging and recruiting under-represented populations.

LSGRs also run the risk of creating group harms. Beyond individual re-identification, there is a concern that through aggregating a sufficient amount of genetic information, and allowing it to be compared to other available databases, LSGRs may permit inferences about groups of people that could be considered harmful on a number of levels. First, there is a risk that genetic information could be mobilized to stigmatize or discriminate against individuals due to their perceived membership in a particular group. Often described as a "group-mediated harm to individuals," this kind of harm can arise in situations when a group is associated with increased genetic risk for having a particularly stigmatizing disease or trait (Hausman 2007). Genetic information also can cause harms to groups themselves where such groups have "structures, leadership, causal capacities, and interests that are distinct from and not reducible to the interests of their members" (Hausman 2008). An evolutionary genetics study reporting migration patterns, for example, could present results that differ from group lore thereby undermining the group but not necessarily harming its members. There are many ways in which this kind of group harms can be expressed, including loss of status in the majority society, self-stigmatization, and dignitary harms to the community (Freeman et al. 2006).

LSGRs pose a particular risk of creating both kinds of group harms because even though data contained in genomic repositories are not associated with personal information, racial and ethnic information is often retained (Hausman 2008). Furthermore, research has made it possible to infer ancestry about a given individual with high reliability, particularly when that individual is from a structured group whose genetic material has been relatively isolated. This means that as genomic data is shared widely, research might produce associations between racial or ethnic groups, and certain traits or medical predispositions. One such example arose in New Zealand in 2006, when researchers reported a variant of the "warrior gene"—associated with traits such as aggression, violence, and impulsivity—as being "strikingly overrepresented" in New Zealand Māori. A lead researcher was quoted as saying that "obviously" the findings meant that Māori men were "going to be more aggressive and violent and more likely to get involved in risk-taking behavior like gambling." (AAP 2006). The claim generated widespread media attention, and led to immediate opposition from Māori and other commentators (Crampton and Parkin 2007).

The fact that certain population groups can have higher frequencies of certain genotypes based on historical patterns of migration, isolation, and other features of population genetics warrants vigilance about the potential for group-mediated harms from genetic research (Hartl and Clark 2007). Even though the individual participants might have agreed to take part in research, current models of informed consent and promises of privacy do not offer protection from these kinds of group-mediated harms. Because of this, LSGRs present wider-ranging threats than those raised by typical research.

Given these concerns, the question is whether or how policy-makers should impose governance structures on LSGRs to minimize risks to groups. To date, there has been some consideration of group harms, at least in the context of the NIH GWAS data sharing policy which required data access committees (DACs) to ensure that proposed research did not pose a risk of creating group harms. It is not clear whether that policy has been effective, and the more recent NIH genomic data sharing policy has dropped concerns about group harms entirely. While a formal review body might not be necessary, other governance options might mitigate worries about group harms. LSGRs could consider requiring that researchers seeking access to data agree to specific limits on data usage when conducting analyses with sensitive data (e.g., race, ethnicity, geography). Alternatively, researchers could stipulate that their results will not unduly impact any specific group in a foreseeably adverse way, placing the burden on the investigator to consider the ramifications of their findings.

6 Conclusion

The capacity to utilize big data represents a substantial shift in the research landscape; our ability to collect, store, share and aggregate data in such expansive ways is a monumental opportunity, but will surely also present significant ethical challenges. While existing policies and procedures may need to be modified to better protect subjects, some scholars have gone further, suggesting that fundamentally new standards of practice should be developed to deal with the unique ethical concerns created by LSGRs (Gymrek et al. 2013). Our analysis suggests, however, that caution is warranted before any major policies are implemented. Much attention has been directed at privacy concerns raised by LSGRs, but perhaps for the wrong reasons, and perhaps at the expense of other relevant concerns. We do not think that there is yet sufficient evidence to motivate enactment of major policy changes in order to safeguard welfare interests, although there might be some stronger reasons to worry about subjects' non-welfare interests. We also believe that LSGRs raise genuine concerns about autonomy and justice. Big data research, and LSGRs in particular, have the potential to radically advance our understanding of human disease. While these new research resources raise important ethical concerns, any policies implemented concerning LSGRs should be carefully tailored to ensure that research is not unduly burdened.

References

AAP. (2006, August 8). Warrior Gene" Blamed for Maori Violence. National Nine News.

Bollier, D., & Firestone, C. M. (2010). *The promise and peril of big data*. Washington, DC: Aspen Institute.

Broadstock, M., Michie, S., & Marteau, T. (2000). Psychological consequences of predictive genetic testing: A systematic review. *European Journal of Human Genetics, 8*(10), 731–738.

Bustamante, C. D., Francisco, M., & Burchard, E. G. (2011). Genomics for the world. *Nature, 475*(7355), 163–165.

Chen, D. T., Rosenstein, D. L., Muthappan, P. G., Hilsenbeck, S. G., Miller, F. G., Emanuel, E. J., et al. (2005). Research with stored biological samples: What do research participants want? *Archives of Internal Medicine, 165*, 652–655.

Citizens' Council on Health Care. (2009). State by state government newborn blood & baby DNA retention practices. Retrieved at http://www.cchfreedom.org/pdf/50_States-Newborn_

Claes, P., Hill, H., & Shriver, M. D. (2014). Toward DNA-based facial composites: Preliminary results and validation. *Forensic Science International: Genetics, 13*, 208–216.

Crampton, P., & Parkin, C. (2007). Warrior genes and risk-taking science. *New Zealand Medical Journal, 120*, U2439.

Dunn, C. K. (2012). Protecting the silent third party: The need for legislative reform with respect to informed consent and research on human biological materials. *Charleston Law Review, 6*, 635–684.

Eiseman, E. (2000). Stored tissue samples: An inventory of sources in the United States. In National Bioethics Advisory Commission (NBAC), *Research involving human biological materials: Ethical issues and policy guidance*. Rockville, Maryland: NBAC.

Federal Policy for the Protection of Human Subjects. (2015). Retrieved at https://www.federalregister.gov/articles/2015/09/08/2015-21756/federal-policy-for-the-protection-of-human-subjects

Freeman, W. M., Romero, F. C., & Kanade, S. (2006). Community consultation to assess and minimize group harms. In E. A. Bankert & R. J. Amdur (Eds.), *Institutional review board management and function* (2nd ed.). Sunderland, MA: Jones and Bartlett.

Geetter, J. S. (2011). Another man's treasure: The promise and pitfalls of leveraging existing biomedical assets for future use. *Journal of Health and Life Science Law, 4*, 1–104.

Genetic Information Non-Discrimination Act Charges. (2014). Retrieved at http://www.eeoc.gov/eeoc/statistics/enforcement/genetic.cfm

Genomic Data Sharing. (2014, August 27). Retrieved at https://gds.nih.gov/

Grady, C., Eckstein, L., Berkman, B. E., Brock, D., Cook-Deegan, R., Fullerton, S. M., et al. (2015). Broad consent for research with biological samples: Workshop conclusions. *American Journal of Bioethics, 15*(9), 34–42.

Green, R. C., Roberts, J. S., Cupples, L. A., Relkin, N. R., Whitehouse, P. J., Brown, T., & Farrer, L. A. (2009). Disclosure of APOE genotype for risk of Alzheimer's disease. *New England Journal of Medicine, 361*(3), 245–254.

Gymrek, M., McGuire, A. L., Golan, D., Halperin, E., & Erlich, Y. (2013). Identifying personal genomes by surname inference. *Science, 339*(6117), 321–324.

Haga, S. B. (2010). Impact of limited population diversity of genome-wide association studies. *Genetics in Medicine, 12*(2), 81–84.

Hartl, D. L., & Clark, A. G. (2007). *Principles of population genetics* (4th ed.). Sunderland, MA: Sinauer Associates.

Hausman, D. M. (2007). Group risks, risks to groups, and group engagement in genetics research. *Kennedy Institute of Ethics Journal, 17*, 351–369.

Hausman, D. (2008). Protecting groups from genetic research. *Bioethics, 22*(3), 157–165.

Heshka, J. T., Palleschi, C., Howley, H., Wilson, B., & Wells, P. S. (2008). A systematic review of perceived risks, psychological and behavioral impacts of genetic testing. *Genetics in Medicine, 10*(1), 19–32.

Homer, N., Szelinger, S., Redman, M., Duggan, D., Tembe, W., Muehling, J., & Craig, D. W. (2008). Resolving individuals contributing trace amounts of DNA to highly complex mixtures using high-density SNP genotyping microarrays. *PLoS Genetics, 4*(8), e1000167.

Hudson, K. L., Rothenberg, K. H., Andrews, L. B., Kahn, M. E., & Collins, F. S. (1995). Genetic discrimination and health insurance: An urgent need for reform". *Science, 270*(5235), 391–393.

International Human Genome Sequencing Consortium. (2004). Finishing the euchromatic sequence of the human genome. *Nature, 431*(7011), 931–945.

Joly, Y., Feze, I. N., & Simard, J. (2013). Genetic discrimination and life insurance: A systematic review of the evidence. *BMC Medicine, 11*, 25–40.

Lane, J., Stodden, V., Bender, S., & Nissenbaum, H. (Eds.). (2014). *Privacy, big data, and the public good: Frameworks for engagement*. Cambridge: Cambridge University Press.

McKinnon, W., Banks, K. C., Skelly, J., Kohlmann, W., Bennett, R., Shannon, K., & Wood, M. (2009). Survey of unaffected BRCA and mismatch repair (MMR) mutation positive individuals. *Familial Cancer, 8*(4), 363–369.

Meiser, B. (2005). Psychological impact of genetic testing for cancer susceptibility: An update of the literature. *Psycho-Oncology, 14*, 1060–1074.

National Science Foundation. (2010). Core techniques and technologies for advancing big data science and engineering program solicitation. Retrieved at http://www.nsf.gov/pubs/2012/nsf12499/nsf12499.htm

Olson, J. (2014, January 14). Minnesota must destroy 1 million newborn blood samples. *Star Tribune*.

Otlowski, M., Taylor, S., & Bombard, Y. (2012). Genetic discrimination: International perspectives. *Annual Review of Genomics and Human Genetics, 13*, 433–454.

Peters, S. A., Laham, S. M., Pachter, N., & Winship, I. M. (2014). The future in clinical genetics: Affective forecasting biases in patient and clinician decision making. *Clinical Genetics, 85*(4), 312–317.

Pollitz, K., Peshkin, B. N., Bangit, E., & Lucia, K. (2007). Genetic discrimination in health insurance: current legal protections and industry practices. *INQUIRY: The Journal of Health Care Organization, Provision, and Financing*, *44*(3), 350–368.

Prince, A. E., & Berkman, B. E. (2012). When does an illness begin: Genetic discrimination and disease manifestation. *The Journal of Law, Medicine & Ethics, 40*(3), 655–664.

Roberts, J. S., Christensen, K. D., & Green, R. C. (2011). Using Alzheimer's disease as a model for genetic risk disclosure: Implications for personal genomics. *Clinical Genetics, 80*(5), 407–414.

Roser, M. A. (2009, December 23). State agrees to destroy more than 5 million stored blood samples from newborns. *Statesman*.

Rothstein, M. A. (2010). Is deidentification sufficient to protect health privacy in research? *The American Journal of Bioethics, 10*(9), 3–11.

Rotimi, C. N. (2012). Health disparities in the genomic era: The case for diversifying ethnic representation. *Genome Medicine, 4*(8), 65–68.

Schadt, E. E. (2012). The changing privacy landscape in the era of big data. *Molecular Systems Biology, 8*(1), 612.

Schadt, E. E., Woo, S., & Hao, K. (2012). Bayesian method to predict individual SNP genotypes from gene expression data. *Nature Genetics, 44*(5), 603–608.

Scutti, S. (2014, July 24). The government owns your DNA. What are they doing with It? *Newsweek*.

Suter, S. M. (2014). Did you give the government your baby's DNA? Rethinking consent in newborn Screening. *Minnesota Journal of Law Science and Technology, 15*, 729–790.

Tomlinson, T. (2009). Protection of non-welfare interests in the research uses of archived biological samples. In K. Dierickx & P. Borry (Eds.), *New challenges for biobanks: Ethics, law, governance*. Intersentia: Antwerp.

Tomlinson, T., De Vries, R., Ryan, K., Kim, H. M., Lehpamer, N., & Kim, S. Y. (2015). Moral concerns and the willingness to donate to a research biobank. *Journal of the American Medical Association, 313*(4), 417–419.

Waldo, A. (2010, March 16). The Texas newborn bloodspot saga has reached a sad—and preventable—conclusion. *Genomics Law Report*.

Wendler, D. (2006). One-time general consent for research on biological samples. *British Medical Journal, 332*(7540), 544–547.

Neurotechnological Convergence and "Big Data": A Force-Multiplier Toward Advancing Neuroscience

Diane DiEuliis and James Giordano

1 Introduction: Neuroscience, Convergence, and the Importance of—and Need for—(Big) Data

Historically, neuroscience has employed approaches from the natural sciences to develop "tools-to-theory heuristics" to formulate ever more detailed understanding of the brain. The conjoinment of diverse approaches and disciplines (e.g.—the physical and social sciences), and intentional "technique and technology sharing" has been important to rapid and numerous discoveries and developments in the brain sciences. If and when purposively employed to meet intellectual challenges and/or technical impediments, the capabilities and advancements achievable through such inter-theoretical and technical cooperation become ever more synergistic. This process, advanced integrative scientific convergence (AISC), is not merely a technical sharing, but is a paradigmatic approach to fostering innovative use of knowledge-, skill-, and tool-sets toward de-limiting existing approaches to question/problem resolution, and to developing novel means of addressing and solving such issues. In this way, AISC enables (a) concomitant "tools-to-theory" and "theory-to-tools" heuristics, and (b) translation of heuristics and tools to practice. The AISC model is being increasingly employed within neuroscience to engage and direct computational methods and advancements in yoking new neurotechnologies that assess and affect

---

Force multiplier—a capability that when employed, significantly increases the potential of a force and thereby enhances the probability of successful engagement and outcomes.

D. DiEuliis · J. Giordano (✉)
Department of Neurology and Neuroethics Studies Program, Georgetown University Medical Center, 4000 Reservoir Rd, Washington DC 20057, USA
e-mail: james.giordano@georgetown.edu

D. DiEuliis
e-mail: Diane.DiEuliis@hhs.gov

J. Collmann and S.A. Matei (eds.), *Ethical Reasoning in Big Data*, Computational Social Sciences, DOI 10.1007/978-3-319-28422-4_6

both the structure and function(s) of the brain, and by extension, human cognition, emotion, and behavior (Giordano 2012b; Vaseashta 2012).

The use of AISC in neuroscience is constrained, and to some extent opportuned by difficulties of matching certain types of neurologic information (e.g.—from neuroimaging and/or neurogenetic studies) to databases that are large enough to enable statistically relevant, and meaningful comparative and/or normative inferences. Current and planned uses of AISC approaches in neuroscience are aimed at overcoming these (and perhaps other) constraints (Giordano 2012b; Vaseashta 2012). The paucity or lack of common data (or a dynamic data base and supportive infrastructure) creates difficulties (if not impossibility) of (1) intra-subject, temporal comparisons (e.g.—using amassed time-point and/or lifespan data); (2) small group and cohort inter-subject single and multiple-timepoint comparisons; (3) single subject- and cohort-to-population comparisons; and (4) population-to-cohort and/or subject normative inferences.

Advanced computational capability, as applied to many types of extant neurotechnologies is increasing their inherent potential such that technological advances within individual disciplines of neuroscience are now being potentially force-multiplied: there is growing capability to integrate differing types, and levels of data both within and across disciplines. Current iterations of computational technology and cybersystems maximize storage and retrieval through parallel processing; such applications are scalable and customizable. In addition, the "cybersphere" creates a nexus for the dissemination, exchange and acquisition/engagement of information from science and technology, and is a medium and forum for (iteratively advancing) scientific convergence, integration and socio-cultural influence. Thus, as defined in the context of neuroscience, big data does not merely refer to the accumulation of large volumes of information (although the acquisition and storage of data itself are certainly relevant and important), but includes the handling of large scale and often disparate informational sets, together with new methods of data visualization, assimilation, comparison, syntheses and analyses in order to define and elucidate dynamic, systems' network models of nervous systems—inclusive of the structure and function of the human brain (Giordano 2014a). Big data methods allow some measure of "dimensionality reduction"—the ability to resolve and discover patterns within large, complex data sets. In these ways, big data methods can fortify the capabilities of convergent forms and uses of neurotechnology, and can increase utility in research and its translational applications.

As evidenced by international calls for neuroscientific discovery, such as the European Union's Human Brain Project (https://www.humanbrainproject.eu; European Commission, 2013), the United States' Brain Research through Advancing Innovative Neurotechnologies (BRAIN) initiative (Insel et al. 2013; The White House, 2015), and the behavioral data initiative (https://www.whitehouse.gov/the-press-office/2015/09/15/executive-order-using-behavioral-science-insights-better-serve-american), among others, there are definitive invocations for greater understanding of the structure and functions of the human brain. Such pursuits are primarily oriented toward achieving improvements in human health and aging through increased capacity to treat and/or prevent neuro-psychiatric disorders. However, it is important to recognize and acknowledge the potential to employ both neuroscientifically-derived knowledge and techniques and technologies to address (and perhaps affect) critical global social conditions and relations (Giordano 2012a, b, 2014b, c; Benedikter and Giordano 2012; Giordano and Benedikter 2012). Big data approaches are fundamental to realizing such goals. Development and application of these and related neuroscientific big data endeavors will not only present technical challenges, but will also foster ethical issues, questions and problems focal to the use and interpretation of resultant findings (Giordano 2014a; Benedikter and Giordano 2011). Hence, it will be important—and necessary—to address such issues through pragmatic neuroethical discourse (and engagement) in order to remain apace with advances in neuroscientific progress, the employment of big data approaches, and the effects and meanings that these approaches manifest in the social sphere.

2 Big Data as Neurotechnological Force Multiplier

Important advances in neurotechnologies during the past two decades have enabled both advances in experimental methods, and the translation of new tools to neuro-psychiatric therapeutics. Recent innovations and expansion in optical and photonic systems, and large-scale semiconductor integration have allowed increased capability in neuroimaging, and developments in bio-engineering have led to rapid progress in interventional technologies, inclusive of transcranial electrical and magnetic stimulation (tES/TMS), open- and closed-loop deep brain stimulation (DBS), and other forms of central and peripheral neuroprosthetics. Such efforts could be powerfully augmented—and the information and capability they render optimized—by their concatenation through big data approaches.

Namely, big data computational capability could enable the establishment of a common, accessible database that provides a resource for (1) (raw) data harvesting; (2) data fusion; (3) data integration, functional formulation, and exchange; and (4) broad data access and use. Conceptually, this would create a repository of (multi-factorial and multi-level) data large enough to establish correlative patterns that satisfies (a) methodological validity (b) adequate probabilistic inference, and (c) reliability. But in order to be wholly viable and of genuine value, this database would, in fact, need to be more than a simple repository, and exist as a dynamic—and secure—resource of tools and methods for harvesting (and provenance), quality evaluation (and data retraction if and when quality issues and/or problems are revealed/elucidated), distribution, and sharing (Giordano 2014a). Such an integrated big data system could allow information to be more “legible” to various user and stake-holder communities (e.g.—bioscience, clinical medicine, policy makers). In this way, it could decrease, if not overcome, the informational fragmentation (e.g.—diversity of subjects, protocols and methods), and lexical issues that impede effective inter-disciplinary collaboration within neurosciences, and the communication, and sound assessment and use of neuroscientifically-derived information and tools within particular user groups.

2.1 Exemplar 1: "Big Molecular Data"

In molecular biology, data set gathering to discover/reveal correlative patterns within "molecular noise" represents a well-accepted and commonly used approach. Molecular biologists have routinely assayed genetic and genomic footprints of different neural cell types and tissues in order to gain better understanding of the functions of various brain regions. The elucidation of a 'neural genome' has both been built upon, and allowed a more thorough examination of genes that putatively establish and control development, structure and activities of the nervous system. As well, genetic 'markers' of particular neuropathologies have been identified, as have the presence of genetic factors that could be predispositional to these conditions. Yet, such data analytics, while sound, have provided somewhat static views of molecular signaling cascades at the cellular level that afford only tangential linking of genes, molecules, and cell signals to certain neuronal functions and/or dysfunctions.

An enabling technology, such as next-generation sequencing (NGS), can advance neuro-molecular research to higher level capability by producing big data sets in two important areas: the neuro-transcriptome, and the neuro-epigenome (Maze et al. 2014). Big data computational approaches, as a force multiplier, would enable understanding of the mechanisms by which neurons develop, alter and/or maintain their molecular signatures during information processing and subsequently, function in the generation of systems hierarchies operative in cognition and behavior. Recent studies have identified regulatory epigenetic activity characterizing normal cell development, function, and plasticity, as well as abnormal processes implicated in human disease. The gathering and analysis of neuro-epigenomic data pose novel challenges—but such objectives are now potentially achievable given the relative success of these methods as applied to other cell types. Integration of genome-wide and proteomic approaches will be necessary to fully understand the neuro-epigenome and the extraordinarily complex nature of the human brain. Unlike other cells, neurons have the capability to alter their transcriptome within minutes (Guo et al. 2011).

Therefore, identifying the transcriptome would require the collection of large, high-throughput proteomic data sets so as to compare insights provided by proteomics with those derived from transcriptomic and genomic data, and allow overlaying this information upon activity profiles. While a seemingly daunting endeavor, it is now rationally envisioned in light of big data tools and techniques. These studies could be critical to diagnostic and therapeutic strategies aimed at heterogeneous and genetically distinct central nervous system (CNS) disorders. The demonstrated inherent diversity of individuals' neural development and expression has prompted calls—and the evident need—for a more precise, personalized approach to the treatment of neurological and psychiatric disorders; big data capability may effectively enable such approaches.

2.2 Exemplar 2: "Big Data Connectomics"

The "connectome" refers to a comprehensive brain model that combines neuroimaging (with ultrastructural anatomical precision and resolution) with functional neuronal activity profiles (Lichtman and Denk 2011). Studies are dedicated to mapping mammalian brains on advanced electron microscopy platforms, often utilizing 60 or more beams (Lichtman et al. 2014); these investigations generate vast amounts of digital spatial information at rapid rates. Still, creating a fully comprehensive depiction of the brain's connectivity and activity remains an exceedingly expansive endeavor. As opposed to single electron recording at the level of individual neurons, a multi-electrode recording technology has been adopted that can increase (by orders of magnitude) the number of neurons that can be simultaneously assessed (Ahrens et al. 2013).

Rather than decoding individual activation patterns, the goal is to link the patterned activity of neurons to sensory, cognitive, and motor events, thereby leading to (a) translation of these activity patterns to an enhanced correlation of neural function(s) and cognitive and behavioral states, (b) improved understanding of mechanisms of neuro-psychiatric disease and disorders, and (c) development of fortified or new therapeutics. The ability to envision the connectome will depend and rely upon the capability to establish and sort meaningful patterns from a large volume, and diverse types of data—and to move from "neural similarities, to semantic similarities". This has been described by Cunningham and Yu as moving from visualizing individual, specific "fingerprints", to recognizing "handprints" (2014).

In this light, big data capability also affords force multiplication. Increased computational power and novel computing algorithms allow multivariate analysis of entire populations of neurons, elucidating key representational patterns within and between various forms of neural network activity. Here "dimensionality reduction" serves as a statistical, validated means to reveal neural mechanisms from multivariate population data (Cunningham and Yu 2014). Such dimensionality reduction using big data approaches has been important to further defining neural mechanisms of decision making, learning, memory, and speech (Rigotti et al. 2013; Machens et al. 2010; Cohen and Maunsell 2010).

Future applications may extend novel forms of complex big data instrumentation toward mapping the "transcriptome" to the "connectome" (Marblestone et al. 2014). But the brain "connectome" produces behavior via convergent activity of genes, neural structure and functions, physical constraints and environmental (including sociocultural) effects. Behavior is a complex, highly dimensional, dynamical and relational phenomenon without clear separation of multiple layers (i.e.—types and levels) of overlapping data. Thus, big data applications may provide methods to develop a more insightful view (and understanding) of how embodied brains function within the environments in which they are embedded.

2.3 *Neuroethical Issues, Concerns and Approaches*

Big data approaches can and will fortify the descriptive, definitive and predictive capacities, use and social influence of neuroscientific information. However, the rapidity with which such advances can—and often do—occur tends to outpace that of ethical address. This is evidenced by recent calls for "moratoria" of particular technological experimentation, to allow for proper identification of ethico-legal and social implications. Therefore, given the current and near term momentum of efforts to establish big data platforms operable in neuroscientific research and its translation, we espouse the urgency of—and responsibility for—dedicating equivalent effort(s) in neuroethical analyses and address.

In general, ethical issues that are most relevant to the engagement of big data in neuroscientific research and its translation reflect concerns about inappropriate access, inapt use/misuse, data modification, and "downstream" effects (e.g.—individual and group socio-economic and legal manifestations of accessed, misused or manipulated datasets). Discourse and debates focusing upon these issues bespeak underlying tensions between accessibility versus sanctuary; privacy versus protection, and libertarian sentiments versus calls for control (NB: A complete discussion of the ethics of big data is beyond the scope of this chapter; for detailed examination, see elsewhere, this volume). We do not advocate impeding science in order to allow ethics to gain ground with progress made to date. Instead, we opine that ethics projects must be poised to address current capabilities, be forward-looking, and pragmatically predictive. Employing a simple precautionary principle to govern the pace and direction of scientific effort is not advocated given that the inherent "character" of frontier science is shaped by change. Hence, benefit (s) incurred by the use of cutting-edge science and/or technology are usually proximate, while risks, burdens, and harms tend to arise after a period of time (Giordano 2012a, b, 2014b, c, 2015).

Therefore, a more realistic—and useful—stance is one of preparedness, in which benefits, threats, and vulnerabilities are identified and assessed, and integrative models and methods of science, technology, and ethics are used to target, mitigate and/or counterbalance these risks and maximize specifically defined goods (Giordano 2011a, b, 2012a, b, 2014b, c, 2015). We argue that just as there are force multipliers that advance neuroscientific methods and tools, there is a defensible need to develop neuroethical discourse and action to 'force multiply' the probity of using big data in and for neuroscientific research and its translation. Such discourse must: (1) acknowledge and define the changing neuroscientific capabilities conferred by the use—and/or misuse—of big data approaches; (2) identify those neuroethico-legal and social issues generated by such use and effect(s); and (3) establish methods to address and resolve such issues, questions and problems, in part through both the development of (practice) guidelines, and by informing and contributing to public policy.

This prompts the question of whether existing ethical methods and systems are viable for addressing, analyzing, guiding, directing and governing those ways that big data will be used in and for neuroscientific research and its varied applications. We believe that they are not; at least not so as to fully and satisfactorily address the contingencies likely to be generated by such large scale acquisition, accessibility and utilization of information—and the effects that such information will incur in each and all respective domains of use and meaning (Giordano 2011a, b, 2014a, b, c; Giordano and Olds 2010; Shook and Giordano 2014). It has been claimed that neuroethics may offer a "new way of doing ethics" at least in some regards (Levy 2011; Giordano 2011a, b; Shook and Giordano 2014). This may be true; as a field neuroethics focuses upon how progress and uncertainty are affected by—and affect—developments in other disciplines (including computational science). Furthermore, we have argued, pro philosopher Fritz Jahr, that as science advances and provides new knowledge and capabilities, ethics must also advance, and must address and incorporate such knowledge, and acknowledge the influence of knowledge and scientific and technological capacity upon society (Giordano et al. 2012).

Thus, it will be important to evaluate, revise and/or develop new ethical concepts and systems that could be employed to evaluate and guide decision-making and action, and establish frameworks to execute such ethical engagement (Shook and Giordano 2014; Lanzilao et al. 2013). Toward this end, we have previously proposed that ethical engagement must have *TASKER* properties: it should be *t*emporally- and *t*ask-*a*gile, *s*cientifically and *s*ituationally *k*nowledgeable, and *e*xperientially and *e*thically *r*eceptive, *r*esponsive, and *r*esponsible (Tractenberg et al. 2014). This process is well-aligned with, and supportive of the critical analytic approach (Choudhoury et al. 2009), and a surety framework to address neuroethico-legal and social issues spawned by the use of big data in the brain sciences (Shaneyfelt and Peercy 2012).

3 Conclusions

The employment of big data approaches as "force multiplier" to technological advances in the neurosciences presents an unprecedented opportunity to understand and affect the human brain, human cognition and behavior, and to incur benefits in human health and social conditions. Beyond simple multi-disciplinarity, big data analytics provide new methods and forums for addressing seemingly intractable questions. Big data methods enable the kinds of comparisons necessary to empower the use of neuroscience in a variety of settings and may provide a nexus for the dissemination and exchange of vast (and diverse types of) neuroscientific information (Vaseashta 2012; Giordano 2012b, 2014a). However, such advancement(s) can also incur ethico-legal and social issues. The acquisition, use and analysis of big data can be and/or become problematic to the application of neuroscience and neurotechnology (see, for example, Ioannidis 2005; Gelman and O'Rourke 2014) that can be exacerbated if and when data are employed beyond academic settings, in social (i.e.—legal, economic) and political realms.

To be sure, much of the development and employment of variable scale databanks that that allow for rapid, real-time data collection, analysis, and utilization of big data will occur in socio-economic or political institutions/contexts in which there is considerable pressure to produce actionable—although not necessarily accurate—analyses and interpretations of information. The validity, reliability, and epistemological integrity of these data may not be valued or even perceived to be relevant. Without validity, reliability, and integrity, advances in neuroscience and neurotechnology can be undermined (Ioannidis 2005; Giordano and Benedikter 2012; Gelman and O'Rourke 2014) because the "information" that is disseminated and exchanged is weak or false (Ioannidis 2005; Benjamini and Hechtlinger 2014; Jager and Leek 2014).

Challenges to the epistemological integrity of big data will be amplified if they are not addressed more effectively than those attempts rendered to date. Recent large-scale investments in high throughput basic and translational science agendas, such as the BRAIN initiative, provide considerable impetus and funding to use big data to define and shape the ways that neuroscientific information is incorporated and applied in medicine, public life, and national security and defense programs (Giordano 2014a, b, c).

Therefore, while the possibility of the acquisition, analysis, and/or use of big data may promise some of the potential of the neuroscientific developments, it will be important to assess, analyze, develop, and guide the uses of big data approaches to neuroscientifically-based information that can—and likely will—be engaged. Effectively attending to these contingencies will require (1) pragmatic assessment of the actual capabilities and limits of big data approaches to neuroscience discovery and application(s), (2) open discourse to address the intended and/or unintended outcomes of new knowledge and scientific/technological achievements that may be produced, and (3) recognition of those ways that such outcomes can affect humanity, the human condition, and society—both locally and internationally—on the twenty first century global stage.

Acknowledgements This chapter was supported in part, by funding from Children's Hospital and Clinics Foundation (JG), and an unrestricted research grant from Thync Biotechnologies (JG). Sections of this work were derived from a series of governmental whitepapers produced and edited by the authors (DD, JG) for the Strategic Multilayer Assessment Group of the Joint Staff of the Pentagon, and have been excerpted and used here with permission.

References

Ahrens, M. B., et al. (2013). Whole-brain functional imaging at cellular resolution using light-sheet microscopy. *Nature Methods, 10*, 413–420.

Baltimore, D., Berg, P., Botchan, M., Carroll, D., Charo, R. A., Church, G., et al. (2015). A prudent path forward for genomic engineering and germline gene modification. *Science, 348* (6230), 36–38.

Benedikter, R., & Giordano, J. (2011). The outer and inner transformation of the global sphere through technology: The state of two fields in transition. *New Global Studies, 5*(2).

Benedikter, R., & Giordano, J. (2012). Neurotechnology: New frontiers for European policy. *Pan Euro Network Sci Tech, 3*, 204–207.

Benjamini, Y., & Hechtlinger, Y. (2014) Discussion: An estimate of the science-wise false discovery rate and applications to top medical journals by Jager and Leek. *Biostatistics, 15*(1), 13–6. (discussion 39–45).

Choudhoury, S., Nagel, S. K., & Slaby, J. (2009). Critical neuroscience: Linking neuroscience and society through critical practice. *Biosocieties, 4*(1), 61–77.

Cohen, M. R., & Maunsell, J. H. R. (2010). A neuronal population measure of attention predicts behavioral performance on individual trials. *Journal of Neuroscience, 30*, 15241–15253.

Cunningham, J. P., & Yu, B. M. (2014). Dimensionality reduction for large-scale neural recordings. *Nature Neuroscience, 17*(11), 1500–1509.

European Commission. (2013). *The human brain project*. Retrieved from https://www.humanbrainproject.eu/

Gelman, A., & O'Rourke, K. (2014). Discussion: Difficulties in making inferences about scientific truth from distributions of published p-values. *Biostatistics, 15*(1), 18–45.

Giordano, J. (2011a). Neuroethics-two interacting traditions as a viable meta-ethics? *AJOB-Neuroscience, 3*(1), 23–25.

Giordano, J. (2011b). Neuroethics: Traditions, tasks and values. *Human Prospect, 1*(1), 2–8.

Giordano, J. (2012a). Neurotechnology as deimurgical force: Avoiding Icarus' folly. In J. Giordano (Ed.), *Neurotechnology: Premises, potential and problems* (pp. 1–15). Boca Raton: CRC Press.

Giordano, J. (2012b). Integrative convergence in neuroscience: trajectories, problems and the need for a progressive neurobioethics. In A. Vaseashta, E. Braman & P. Sussman (Eds.), *Technological innovation in sensing and detecting chemical, biological, radiological, nuclear threats and ecological terrorism.* (NATO Science for Peace and Security Series), New York: Springer.

Giordano, J. (2014a). Intersections of "big data", neuroscience and national security: Technical issues and derivative concerns. In H. Cabayan, D. DiEuliis, et al. (Eds.), *A new information paradigm? From genes to "Big Data", and instagrams to persistent surveillance: Implications for national security* (pp. 46–48). Washington, DC: Department of Defense; Strategic Multilayer Assessment Group-Joint Staff/J-3/Pentagon Strategic Studies Group, November, 2014.

Giordano, J. (2014b). The human prospect(s) of neuroscience and neurotechnology: Domains of influence and the necessity—and questions—of neuroethics. *Human Prospect, 4*(1), 1–18.

Giordano, J. (2014c). Neurotechnolgoy, global realtions, and national security: Shifting contexts and neuroethical demands. In J. Giordano (Ed.), *Neurotechnology in national security and defense practical considerations, neuroethical concerns* (pp. 1–10). Boca Raton: CRC Press.

Giordano, J. (2015). A preparatory neuroethical approach to assessing developments in neurotechnology. *AMA J Ethics, 17*(1), 56–61.

Giordano, J., & Benedikter, R. (2012). An early—and necessary—flight of the Owl of Minerva: Neuroscience, neurotechnology, human socio-cultural boundaries, and the importance of neuroethics. *Journal of Evolution and Technology, 22*(1), 14–25.

Giordano, J., Benedikter, R., & Kohls, N. B. (2012). Neuroscience and the importance of a neurobioethics: A reflection upon Fritz Jahr. In A. Muzur & H.-M. Sass (Eds.), *Fritz Jahr and the foundations of integrative bioethics*. Münster, Berlin: LIT Verlag.

Giordano, J., & Olds, J. (2010). On the interfluence of neuroscience, neuroethics and legal and social issues: The need for (N)ELSI. *AJOB-Neuroscience, 2*(2), 13–15.

Guo, J. U., et al. (2011). Neuronal activity modifies the DNA methylation landscape in the adult brain. *Nature Neuroscience, 14*, 1345–1351.

Insel, T. R., Landis, S. C., & Collins, F. S. (2013). The NIH brain initiative. *Science, 340*, 687–688.

Ioannidis, J. P. A. (2005). Why most published research findings are false. *PLoS Medicine, 2*(8), e124.

Jager, L. R., & Leek, J. T. (2014). An estimate of the science-wise false discovery rate and application to the top medical literature. *Biostatistics, 15*(1), 1–12.

Lanzilao, E., Shook, J., Benedikter, R., & Giordano, J. (2013). Advancing neuroscience on the 21st century world stage: The need for and proposed structure of an internationally relevant neuroethics. *Ethics in Biology, Engineering and Medicine, 4*(3), 211–229.

Levy, N. (2011). Neuroethics-a new way of doing ethics. *AJOB-Neuroscience, 2*(2), 3–9.

Lichtman, J. W., & Denk, W. (2011). The big and the small: Challenges of imaging the brain's circuits. *Science, 334*(6056), 618–623.

Lichtman, J. W., Helmstaedter, M., & Sanders, S. (2014). Connectomics at the cutting edge: Challenges and opportunities in high-resolution brain mapping. *Science Webinar Series*, Transcript retrieved at: http://bit.ly/1LsdgSD

Machens, C. K., et al. (2010). Functional, but not anatomical, separation of 'what' and 'when' in prefrontal cortex. *Journal of Neuroscience, 30*, 350–360.

Marblestone, A., Daugharthy, E., Kalhor, R., Peikon, I., Kebschull, J., Shipman, S., et al. (2014). Rosetta brains: A strategy for molecularly-annotated connectomics (Retrived on ArXiv).

Maze, I., et al. (2014). Analytical tools and current challenges in the modern era of neuroepigenomics. *Nature Neuroscience, 17*, 1476–1490.

Rigotti, M., et al. (2013). The importance of mixed selectivity in complex cognitive tasks. *Nature, 497*, 585–590.

Shaneyfelt, W., & Peercy, D. E. (2012). A surety engineerting framework asnd process to address ethical legal and social issues for neurotechnology. In J. Giordano (Ed.), *Neurotechnology: Premises, potential, and problems* (pp. 213–232). Boca Raton: CRC Press.

Shook, J. R., & Giordano, J. (2014). A principled, cosmopolitan neuroethics: Considerations for international relevance. *Philosophy, Ethics, and Humanities in Medicine, 9*, 1.

The White House (2015). *Executive order—using behavioral science insights to better serve the American people*. Retrieved from https://www.whitehouse.gov/the-press-office/2015/09/15/executive-order-using-behavioral-science-insights-better-serve-american

Tractenberg, R., FitzGerald, K. T., & Giordano, J. (2014). Engaging neuroethical issues generated by the use of neurotechnology in national defense: Toward process, methods, and paradigm. In J. Giordano (Ed.), *Neurotechnology in national security and defense practical considerations, neuroethical concerns* (pp. 259–278). Boca Raton: CRC Press.

Vaseashta, A. (2012). The potential utility of advanced sciences convergence: Analytical methods to depict, assess, and forecast trends in neuroscience and neurotechnological developments and uses. In J. Giordano (Ed.), *Neurotechnology: Premises, potential, and problems* (pp. 15–36). Boca Raton: CRC Press.

Data Ethics—Attaining Personal Privacy on the Web

Lisa Singh

1 Introduction

As digital communications continue to increase, people continue to share more and more data, including personal information. As of fall 2015, there were more than 3.2 billion Internet users (Real Time Statistics Project 2015); of these users, 1.5 billion share information on Facebook, 343 million on Google+, 380 million on LinkedIn, and 316 million on Twitter (Smith 2015). Because much of the data shared are publicly accessible, a large opportunity exists for data mining researchers to develop algorithms and methods to support a wide array of analytic services dependent on understanding human preferences. Examples include recommendation systems and customized search tools. On the flip side, the sharing of these large amounts of personal information, some of which are more sensitive in nature, is concerning in the context of personal privacy.

While segments of the population utilize privacy features offered by social media sites, many Internet users do not. Personal demographic information, as well as ideas and thoughts (tweets or messages) that once would have been shared in a more private setting with groups of friends/acquaintances are now accessible to anyone with a computer. If every person and company used these data in ethically responsible ways, then the sharing of so much personal data would be inconsequential. Unfortunately, this is not the case. Individuals, researchers, and companies are using these data beyond their original intent (Tompsett 2005; Hill 2012; Soper 2012; Kramer et al. 2014). The question we have to ask ourselves is—what should we consider reasonable, ethical uses of personal online data?

This chapter begins by discussing some ethically questionable uses of personal data. It then identifies different technologies that are being developed to improve individual privacy on the Internet. The goal of these technologies is to give users

L. Singh (✉)
Department of Computer Science, Georgetown University, Washington, DC 20057, USA
e-mail: Lisa.Singh@georgetown.edu

J. Collmann and S.A. Matei (eds.), *Ethical Reasoning in Big Data*,
Computational Social Sciences, DOI 10.1007/978-3-319-28422-4_7

more control over their data so that the chance of misuse decreases. Finally, this chapter concludes with a discussion about strategies for improving the current situation and suggests the formalization of the field of data ethics to tackle ethical issues specific to the sharing of personal data online.

2 Drawing the Line—Ethically Questionable Studies

Currently, there is no single federal law that adequately regulates the collection and use of personal data from the Internet (Jolly 2015). While guidelines and best practices exist, privacy laws are inconsistent (and sometimes contradictory) across states and dated at the federal level when it comes to limiting use and sharing of personal, behavioral data. At universities, Institutional Review Boards (IRB) have inconsistent standards related to the use of human subject data from the Internet, i.e. ethical uses of available big data about individuals (SACHRP-HHS 2013). Given the inadequate guidelines related to corporate responsibility of personal data and research that uses human behavioral data from the Internet, it is not surprising that a number of ethically questionable uses of data have arisen. We focus on studies from the field of computer science, not because they are more egregious than social science studies, but because they more readily make use of *big data* in their research. This is not surprising since computer scientists can easily manage analysis of large volumes of data obtained from the web. The majority of this section considers two examples, one in cybersecurity and one in data mining. Both of these studies were approved by the researchers respective IRBs. We leave it to you to decide whether or not they should have been.

2.1 Cybersecurity—Planned Attacks and Malicious Software

Some cybersecurity research involves setting up adversarial attacks to better understand insecurities in software. For example, intrusion detection research uses large volumes of network traffic data to generate signatures of potential attacks. Network traffic data contains IP addresses that are not anonymized. However, they are determined using measurement traffic and are not associated with specific individuals (Paxson 2004). Therefore, these analyzes do not violate the privacy of individuals and are important for identifying and preventing different types of cyberattacks. In our viewpoint, this type of research is a good example of ethical big data, cybersecurity research. Of course, we make the assumption that the research follows the federal privacy laws related to access of traffic on computer networks.

There is also a subset of cybersecurity research focused on understanding the proliferation of malware, spam, email harvesting, etc. through the use of botnets. Botnets are a set of compromised machines that can be remotely controlled by an attacker. Because they can grow to be quite large, they are a danger to the Internet community. The most common uses of botnets include Denial-of-Service attacks, spamming, harvesting email addresses, spreading malware, and keylogging (Bacher et al. 2008).

One study used botnets to better understand the "economics of spam" (Kanich et al. 2008). Specifically, they used existing spamming botnets (by infiltrating an existing botnet infrastructure) to understand who and how many people click on spam. They considered two types of spam campaigns, one that propagated malware and one that marketed online pharmaceuticals. To emulate those campaigns and determine click through rates of these forms of spam, the authors created two websites. While the researchers do not actually spread any malware or collect credit card information for their fake pharmaceutical sites, they trick users into believing that they are going to actual sites specified in the spam, e.g. sites where medication can be purchased. Once users click to checkout on one of the websites, an error message is given—no additional personal information is obtained. Neither during or after the process are users informed that they are participating in a study. This could be viewed as a study that manipulates users without their expressed consent. The authors did have IRB approval for this study on the grounds that the authors were not increasing the amount of spam the users were receiving or increasing harm to the users.

While we focus here on only two examples, Burstein has an excellent discussion of legal and ethical approaches for conducting cybersecurity research (Burstein 2008).

2.2 *Personalized Data Mining*

Numerous success stories involving the use of big data in conjunction with machine learning and data mining have lead to improvements in healthcare, political campaigning, crime prevention, and customer service to name a few (Siege 2013). Unfortunately, there are also examples of researchers and companies using data they have without considering individual privacy.

Many companies use customer purchasing data to send targeted advertising to their customers. While in principle, using internal customer information in this way does not violate any privacy laws, the targeting itself can be unethical. One well known example is Target's marketing of pregnancy/baby related products (Hill 2012; Duhigg 2012). Target determined that when significant milestones occur in people's lives, they are more open to changing their purchasing habits, i.e. switching stores and/or products. Once they make the change, they tend to be loyal customers. Given this knowledge, one campaign focused on the life changing moment of having a baby. Target was able to combine demographic data with purchasing data for approximately 25 products to identify women who were pregnant. Their analytics were precise enough to predict the baby's approximate due date and then market products based on that inference. While on the surface, learning something private about your customers may seem like good customer mining, in this case, Target chose to market coupons to anyone woman they predicted to be pregnant, including teenagers. In one case, they marketed to a teenager whose parents were unaware that she was pregnant. By most people's standards, this is not ethically appropriate. Even if it was a mistake, we need to make sure that companies consider it their obligation to conduct analyses that pass common sense and basic ethics tests.

The final study we consider in this subsection is one conducted by researchers at Facebook and Cornell University (Kramer et al. 2014). To better understand the effects of reading positive and negative articles, these researchers ran a emotional contagion experiment. During a one week period in 2012, Facebook intentionally changed the news feed of over 650,000 random English-speaking Facebook users. For one group they posted news deemed to be more positive. For the other, the top posts presented were more negative. The researchers then measured whether this adjusting of content had an effect on the emotional status updates of the study users. They found that it did (but the statistical significance was small).

The researchers did not obtain consent from the users in the study. Facebook chose to manipulate people's emotions without their consent. The Cornell IRB indicated that the study was exempt from IRB approval since the faculty and student involved did not directly engage with the user data, but instead only had access to the results. The Facebook Terms of Use and Data Use Policy also do not indicate that these types of psychology experiments may be conducted on users of their site. It is a general consent form that does not have the same depth as an informed content document would.

There has obviously been a fair amount of discussion about the ethics of this study (Gorski 2014; Waldman 2014; Chambers 2014). Companies change what we see on their sites all the time. What is concerning about this study is that they chose to knowingly make a subset of their users less happy without telling them. Neither of these studies rise to the negligence of some of the unethical medical studies we have seen, but it is a preview of the types of studies we may see if we do not develop adequate guidelines for human behavioral studies involving big data.

2.3 Online Tracking of Users

A decade ago, online tracking was conducted using simple "cookies" that recorded when a user visited a website and what they searched for on a website. Now, more advanced tools can not only track browsing behavior, but can link that behavior to personal user data including location, demographic data, and even health data (Valentino-Devries 2010; Olsen 2002).

Even more disturbing is that advertisers are paying companies to track people as they use the Internet to better understand what websites they visit or applications they use (EPIC 2015). While this information could be used for targeting ads, it could also be used to target ads in a biased way, that leverages demographic data to `adjust' prices of the same item for different subgroups. In 2010, the Wall Street Journal conducted a study of tracking technologies and found that the top 50 websites installed an average of 64 pieces of tracking technology on their visitors computers, generally, without any warning. Life insurance companies find policies to advertise that fit a user's demographics; health and drug companies map advertising to health terms users are searching for and health related sites they are visiting; and at least one company with social network data is selling it to companies to understand people's creditworthiness—people who are responsible credit users will 'hang out' with other responsible users (Angwin 2010).

Obviously, this type of data collection can violate different Fair Information Practices (EPIC 2015). Users do not know that companies are doing this tracking, they do not know the specifics of the data that is being collected about them, they do not know how the data will be used or with whom it will be shared, and they do not know how accurate it actually is or what inferences are being made with it. Because of these types of tracking software, users cannot make informed judgments about what to share. It also limits their ability to control their data. A need exists to regulate what can be collected and for how long. A need exists for users to be informed about the data values being stored about them and the data values being infered about them.

3 Technologies Being Developed to Improve Privacy on the Web

Most people in the US have a web presence. Obviously, not using the Internet is the safest option, but an unrealistic one in this technological age. While it is unclear how we can improve the ethics of those using large-scale human behavioral data, there are tools available that can make users more anonymous on the Internet and/or can help them better understand the data that companies have about them or that is publicly available on the web. In this section, we consider the types of technologies that either exist or are being developed to help users improve their privacy or better understand their data.

Computer science researchers investigate ways to protect the privacy of user data. In data mining, they focus on privacy preserving methods that hide identifiable information about the user while still maintaining the utility of the data for statistical analysis and data mining. Some recent methods that give companies ways to share or use personally identifiable data without knowing the identity of the individuals include differential privacy for statistical databases (Dwork 2008), anonymization techniques for relational and network data (Zhou et al. 2008; Samarati and Sweeney 1998; Machanavajjhala et al. 2007; Li et al. 2007), approaches for informing users about their Internet data (Singh et al. 2015; Irani et al. 2009), giving users privacy scores to assess their level of vulnerability (Singh et al. 2015; Luo et al. 2009; Gundecha et al. 2011), and prototypes of user controlled identity management systems (Fang and LeFevre 2010; Lucas and Borisov 2008; Luo et al. 2009). While progress is being made, most of these methods are still academic and have not been integrated into real world systems. We surmise part of the issue is that users acceptance of these practices. They are not outraged enough about these practices, so companies have not made data privacy a priority.

Tools that attempt to give users information about their public profile are being developed. One tool that researchers at Georgetown University are working on is part of the Web Footprint project (Singh et al. 2015). This tool constructs *web footprints* of different users by combining publicly accessible information from various online services such as social media sites, micro-blogging sites, data aggregation sites, and search engines about the users. It essentially emulates an adversary searching for publicly available information about a user and has a goal of informing users about data that can be discovered about them. It also recommends the removal of pieces of data that were instrumental in improving the probability of linking and identifying more data attributes. For example, a person's place of work is usually indicative of his or her home state; similarly, a user's home telephone number can be used to infer his or her city. To ensure that adversaries do not use this software, the software only allows authenticated users to check only their names. This software is in prototype phase, but tools like this will be instrumental in helping users understand their public profiles and make adjustments if they choose to.

Tools and best practices also exist to help users reduce the level of web tracking of companies. Here we highlight a few that have been shown to be effective:

- Do not post private information on social media sites. If you choose to, make sure your privacy settings are not set to allow public access.
- Set your browser to not accept cookies from sites you have not visited or sites you do not want to track you. If you do not want to be tracked by the browser itself, some browsers like Chrome have an option for this (incognito mode).
- Do not respond to spam or click on links to sites you do not know.
- Install an ad blocker. This will improve your computer's performance and will reduce data about your click thru habits.
- Referrer data is information that is collected by the previous site you visited. Install a tool to remove referrer data so that other sites cannot access it.
- Encrypt your email so that it can not be viewed by others.

To find other helpful tips, we refer you to (Schmitz 2013; McCandlish 2002; Neagu 2014). These articles describe different types of attacks and possible ways to deal with them.

4 The Pillars of Data Ethics

A survey by Pew Research in 2014 (Madden 2014) showed that most Americans are concerned about data privacy. Over 90 % of adults surveyed agreed that consumers are no longer in control of how personal data is collected and used by companies. Just under 90 % of adult respondents believe it would be hard to remove inaccurate information about them that is online and 80 % that use social media sites are concerned about access to their personal data by advertisers or other businesses. Yet, even with all this public concern, consumers allow companies to do whatever they want with their behavioral data. The idea of not getting free search, Gmail, or Facebook is considered a greater evil than giving these companies free reign on personal data. Until users change their position and hold companies to higher standards, companies will exploit these data as much as they can.

We are at an interesting time—a time when companies and researchers are using technology to drive the understanding of human behavior at an incredible pace. We need to pause and think about what is happening. We need to take back our personal data rights. We need to enforce ethically appropriate use of personal data. It is not big data or big data technologies that are privacy invaders. It is the way people use big data technologies that is invasive and unethical. Here we propose different strategies for improving the current situation.

Regulation. Users cannot regulate companies. Governments need to step up and develop sound regulation about how much and for how long personal data can be collected. Companies should learn from their customer data, but they should do so in a responsible way. If companies cannot do it themselves, then regulations need to be developed.

Data ethics standards. Data ethics standards related to the use of big data need to be developed. There are no safeguards for consumers right now. Because data ethics are complicated, we need a lot of discussion and debate. It is a ripe area for a new discipline to address the complexities that are arising. Data ethics differs from other forms of ethics and needs fresh eyes assessing the moral implications of sharing different types of data.

Catalog of personal data. Individuals need a way to see the data fields a company maintains about them. One way to do this is to setup a mechanism for user to maintain a catalog of the different personal data companies have access to. Users should also have access to new data that is inferred from the original data the company has about them. This is important because the inferences may be inaccurate and users do not currently have a way of knowing that these inaccuracies exist.

Correct inaccurate data. Not all data, original or inferred, is accurate. Therefore, a straightforward mechanism to correct inaccurate data that companies have is important. We can imagine a registry where users have a list of companies that have different data about them and the registry allows users to request companies remove certain data that is too sensitive and/or update it if it is incorrect.

All of these strategies would improve the current situation and allow users to feel more in control of their personal behavioral data.

5 Concluding Thoughts

Companies have a choice about how they use customer behavioral data. Users also have a choice about what behavioral data they share. Unfortunately, the cultural acceptance of publicly sharing personal information on different social media sites and of allowing companies to collect behavioral data without limitations on what they collect or how long they maintain the data for is troubling. The public has been trained that once the data is collected by a company, the company can use it for purposes beyond the original intended use.

This chapter highlighted a number of cases when companies and/or researchers stepped over the boundary of ethically reasonable uses of the data they had. It also highlighted studies that manipulated individuals online without their expressed consent. Finally, it described some technologies that could improve the level of user privacy on the Internet and recommended strategies to help users gain more control over their data.

Every form a user fills out, every click a user makes on a website, every comment or recommendation a user posts about a product, every decision a user makes online is a new data point that is being used by companies and researchers to better understand and potentially infer human behavior. The time has come to pause and debate online privacy and ethical uses of large-scale human behavioral data. The time has come to develop guidelines and regulations that protect users while still allowing companies and researchers the ability to advance knowledge about human behavior in responsible ways. The time has come to take control of our personal data.

Acknowledgments This work was supported by NSF CNS-1223825 and NSF IIS-1522745. The opinions and findings described in this paper are those of the author and do not necessarily reflect the views of the National Science Foundation.

References

Angwin, J. (2010). The web's new gold mine: Your secrets. Retrieved November 01, 2015, from http://www.wsj.com/articles/SB10001424052748703940904575395073512989404.

Bacher, P., Holz, T., Kotter, M., & Wicherski, G. (2008). Know your enemy: Tracking botnets. Retrieved November 01, 2015, from https://www.honeynet.org/papers/bots.

Burstein, A. (2008). Conducting cybersecurity research legally and ethically. In *Usenix workshop on large-scale exploits and emergent threats*.

Chambers, C. (2014). Facebook fiasco was Cornell's study of 'emotional contagion' an ethics breach? Retrieved November 01, 2015, from http://www.theguardian.com/science/head-quarters/2014/jul/01/facebook-cornell-study-emotional-contagion-ethics-breach.

Duhigg, C. (2012). How companies learn your secrets. Retrieved November 01, 2015, from http://www.nytimes.com/2012/02/19/magazine/shopping-habits.html.

Dwork, C. (2008). Differential privacy: A survey of results. In *Theory and applications of models of computation* (pp. 1–19). Springer.

EPIC. (2015). Online tracking and behavioral profiling. Retrieved November 01, 2015, from http://epic.org/privacy/consumer/online_tracking_and_behavioral.html.

Fang, L., & LeFevre, K. (2010). Privacy wizards for social networking sites. In *ACM world wide web conference (www)*.

Gorski, D. (2014). Did Facebook and PNAS violate human research protections in an unethical experiment? Retrieved November 01, 2015, from https://www.sciencebasedmedicine.org/did-facebook-and-pnas-violate-human-research-protections-in-an-unethical-experiment/.

Gundecha, P., Barbier, G., & Liu, H. (2011). Exploiting vulnerability to secure user privacy on a social networking site. In *Proceedings of the 17th ACM SIGKDD International Conference on Knowledge Discovery and Data Mining* (pp. 511–519). ACM.

Hill, K. (2012). How target figured out a teen girl was pregnant before her father did. Retrieved November 01, 2015, from http://www.forbes.com/sites/kashmirhill/2012/02/16/how-target-figured-out-a-teen-girl-was-pregnant-before-her-father-did/.

Irani, D., Webb, S., Li, K., & Pu, C. (2009). Large online social footprints—An emerging threat. In *International conference on computational science and engineering*.

Jolly, I. (2015). Data protection in united states: Overview. Retrieved November 01, 2015, from http://us.practicallaw.com/6-502-0467.

Kanich, C., Kreibich, C., Levchenko, K., Enright, B., Voelker, G. M., Paxson, V., Savage, S. et al. (2008, October). Spamalytics: An empirical analysis of spam marketing conversion. In *Acm conference on computer and communications security* (pp. 3–14). Alexandria, Virginia, USA.

Kramer, A., Guillory, J., & Hancock, J. (2014). Experimental evidence of massive-scale emotional contagion through social networks. *Proceedings of the National Academy of Science, 111*(42), 8788–8790.

Li, N., Li, T., & Venkatasubramanian, S. (2007, April). t-closeness: Privacy beyond k- anonymity and l-diversity. In *Proceedings of the International Conference on Data Engineering (ICDE).*

Lucas, M.M., & Borisov, N. (2008). FlyByNight: Mitigating the privacy risks of social networking. In *Acm Workshop on Privacy in the Electronic Society (WPES).*

Luo, W., Xie, Q., & Hengartner, U. (2009). FaceCloak: An architecture for user privacy on social networking sites. In *International Conference on Computational Science 13 and Engineering (CSE).*

Machanavajjhala, A., Gehrke, J., & Kifer, D. (2007). l-diversity: Privacy beyond k- anonymity. *ACM Transactions on Knowledge Discovery from Data, 1*(1).

Madden, M. (2014). Public perceptions of privacy and security in the post-snowden era. Retrieved November 01, 2015, from http://www.pewinternet.org/2014/11/12/public-privacy-perceptions/.

McCandlish, S. (2002). EFF's top 12 ways to protect your online privacy. Retrieved November 01, 2015, from https://www.eff.org/wp/effs-top-12-ways-protect-your-online-privacy/.

Neagu, A. (2014). 11 steps to dramatically improve your online privacy in less than 1 hour. Retrieved November 01, 2015, from https://heimdalsecurity.com/blog/online-privacy-essential-guide/.

Olsen, S. (2002). Nearly undetectable tracking device raises concern. Retrieved November 01, 2015, from http://www.cnet.com/news/nearly-undetectable-tracking-device-raises-concern/.

Paxson, V. (2004). Strategies for sound internet measurement. In *Acm Sigcomm Conference on Internet Measurement*. New York, USA: ACM.

Real Time Statistics Project. (2015). Internet live statistics. Retrieved November 01, 2015, from http://www.internetlivestats.com/.

SACHRP-HHS. (2013). Considerations and recommendations concerning internet research and human subjects research regulations, with revisions. Retrieved November 01, 2015, from http://www.hhs.gov/ohrp/sachrp/mtgings/2013%20March%20Mtg/internet_research.pdf.

Samarati, P., & Sweeney, L. (1998). Protecting privacy when disclosing information k-anonymity and its enforcement through generalization and suppression. In *Proceedings of the IEEE Symposium on Research in Security and Privacy.*

Schmitz, D.T. (2013). 5 ways to improve your privacy online. Retrieved November 01, 2015, from http://www.technewsworld.com/story/78590.html.

Siege, E. (2013). *Predictive analytics: The power to predict who will click, buy, lie, or die*. John Wiley & Sons.

Singh, L., Yang, H., Sherr, M., Hian-Cheong, A., Tian, K., Zhu, J., Zhang, S. et al. (2015). Public information exposure detection: Helping users understand their web footprints. In *International Conference on Advances in Social Networks Analysis and Mining (asonam).* Paris, France.

Smith, C. (2015). How many people use 950+ of the top social media, apps, and digital services? Retrieved November 01, 2015, from http://expandedramblings.com/index.php/resource-how-many-people-use-the-top-social-media/.

Soper, D. (2012, April). Is human mobility tracking a good idea? *Communications of ACM*, *55*(4), 35–37.

Tompsett, B. (2005). Identity theft in an onlineworld. *Computer Law Security Report*, *21*(2).

Valentino-Devries, J. (2010). How to avoid the prying eyes. Retrieved November 01, 2015, from http://www.wsj.com/articles/SB10001424052748703467304575383203092034876.

Waldman, K. (2014). Facebook's unethical experiment. Retrieved November 01, 2015, from http://www.slate.com/articles/health_and_science/science/2014/06/facebook_unethical_experiment_it_made_news_feeds_happier_or_sadder_to_manipulate.html.

Zhou, B., Pei, J., & Luk, W. (2008, December). A brief survey on anonymization techniques for privacy preserving publishing of social network data. *ACM SIGKDD Explorations Newsletter*, *10*(2).

Part III
Institutionalizing Ethical Reasoning About Big Data

Technology for Privacy Assurance

J.C. Smart

1 Introduction

Two pillars of a democratic society—Security and Liberty—are challenged by the post-9/11 world: How can an open democracy sustain the former without infringing on the latter? In our new "Big Data" era, a government's ability to collect, process, analyze, and share volumes of information is commonly regarded as central to its national security and its public safety. But these needs, driven by a desire to detect threats and reduce risk to the aggregate population increasingly have been placed in conflict with the constitutional protections of individual liberties.

Current public opinion often frames this tension as a tradeoff, balancing the sacrifice of some liberties against real or perceived gains in security and safety (Center for Strategic and International Studies 2014; Gilmore 2014; Campos 2014). A decade and a half later, no end to this debate is in sight. But the presentation here posits that security/safety and liberty are not mutually exclusive. Rather, it advocates a paradigm that enables both to be achieved simultaneously, through the careful application of policy and modern technology (Smart 2011). This concept and the prescribed implementation approach is referred here as Privacy Assurance.

2 Information Sharing

The sharing of information across legal and jurisdictional boundaries enables new analytic opportunities. From a national security perspective, witness how the 9/11-hijackers were not only connected via airline data and other transactional records, but in at least two cases by threat information already maintained by the

J.C. Smart (✉)
Georgetown University, Washington, D.C., USA
e-mail: smart@georegtown.edu

J. Collmann and S.A. Matei (eds.), *Ethical Reasoning in Big Data*,
Computational Social Sciences, DOI 10.1007/978-3-319-28422-4_8

U.S. Intelligence Community. In the public safety context, HIV spreads between individuals who increasingly receive care and treatment across many jurisdictional boundaries that span where they live, work, and socialize. The new spectrum of contemporary analytic techniques is often popularized as "connecting the dots." But localized information "stovepipes" maintained by individual organizations often are not sufficiently rich in their content to discern the complex network of associations and connections across multiple jurisdictions that realistically describe contemporary threats or societal risks. In contrast, such patterns often are quickly revealed when these otherwise disparate information sources can be merged and analyzed in aggregate.

Unfortunately, the merging of information sources can quickly exceed the respective policies and authorities of participating organizations, creating the new tensions to individual liberties and personal privacy. Alternatively stated, while it often may be in the best interests of single organizations spanning various legal and jurisdictional boundaries to share information, there may not be adequate trust among the participants, or authority from the citizenry under whom they serve, to allow such sharing. This reluctance or mistrust can arise from the fear of misuse with insufficient oversight, fear of the exposure of sensitive information, sources, and methods, or the increased risk of unintentional exposure. Trust and fear issues aside, privacy policy in the United States today mandates data minimization—to wit, that civilian agencies should only collect personally identifying information (PII) that is directly relevant and necessary to accomplish the specified purpose of its collection; only retain PII for as long as is necessary to fulfill the specified purpose; and only share data with other agencies when compatible with the purpose for which it was collected. Moreover, U.S. citizens are afforded constitutional assurance to be "secure in their persons, houses, papers, and effects, against unreasonable searches." Is it possible to achieve national security and public safety goals without eroding such fundamental privacy rights?

The paradigm advocated here takes the Fourth Amendment to the United States Constitution as a basic system requirement. Within this framework from a national security perspective, U.S. law defines "reasonable suspicion" as the standard of law, based on specific and articulable facts and inferences, under which a person may be regarded as being engaged in criminal activities, having been engaged in such activity, or about to be engaged in it. An analog can be readily devised for the public safety sector with "reasonable concern" as the rubric, based on specific and articulable facts and inferences, under which a person may be regarded as being engaged, having been engaged, or about to be engaged in behavior that exposes the public to undue risk.

Reasonable suspicion is the basis for investigatory stops by the police and requires less evidence than probable cause, the legal requirement for arrests and warrants. Analogously, reasonable concern is a basis for required public health organization reporting (e.g. detection of an highly infectious disease) versus higher thresholds requiring quarantine, mandatory evacuation, imposition of marshal law, etc. Reasonable suspicion or reasonable concern are evaluated using the "reasonable person" standard, in which an official (e.g. police officer or public health officer) in the same circumstances could reasonably believe a person has been, is, or is about to be engaged in an activity that seriously jeopardizes the public's security and/or safety.

Such suspicion or concern cannot simply be based on a hunch. A combination of particular facts, even if each is individually innocuous, can form the reasonable suspicion or reasonable concern. This is pivotal to Constitutional law enforcement and to the method for assuring privacy that is laid out below. It describes how reasonable suspicion (concern) can be ascertained from multiple information sources without resorting to unreasonable search. Unreasonable search is interpreted here as any type of investigative process that would reveal information that a reasonable person would regard as private, prior to the establishment of reasonable suspicion/concern or probable cause—and thus protected.

3 Privacy Assurance

So how can reasonable suspicion (concern) be responsibly ascertained from multiple information sources without resorting to unreasonable search, and thus jeopardizing individual privacy? One approach commonly attempted today is the use of anonymization. That is, all discerning PII is removed, sometimes replaced with statistical results versus actual data, sharing only information that is non-identifiable. Unfortunately, in the new "Big Data" era, true anonymization becomes increasingly difficult at increasing scale, as relationships previously hidden among the enormous data complexity can be revealed as processing of larger and larger data volumes from greater numbers of sources continues to grow. Alternatively, anonymization techniques that truly are effective at scale often dramatically reduce the value of the information being exchanged and its ability to enable actionable outcomes. This is particularly apparent in public health applications where the goals are ultimately to genuinely improve the condition of individuals, versus simply a statistical awareness of an aggregate population's inevitable plight.

The privacy approach advocated here posits the existence of a "Black Box." In this context, a Black Box is a physical (or logical) device whose contents are beyond reach: that is, its contents can *never* be examined. The device is specifically engineered so that the information it is fed cannot be revealed to anyone under any circumstances, regardless of authorization, executive privilege, court order, vandalism, or deliberate attack. Information can flow into the Black Box, but once it resides within its boundaries, it can never be accessed. For all practical purposes, the Black Box is considered an impenetrable information container.

Total impenetrability, however, implies a theoretical extreme that likely would be difficult to achieve, or even more important, to verify or accept in the negative. Consequently, this paper treats impenetrability as the condition in which there exist no known exploitable vulnerabilities that would enable access to the contents of the Black Box. While vulnerabilities may exist, an impenetrable Black Box is one

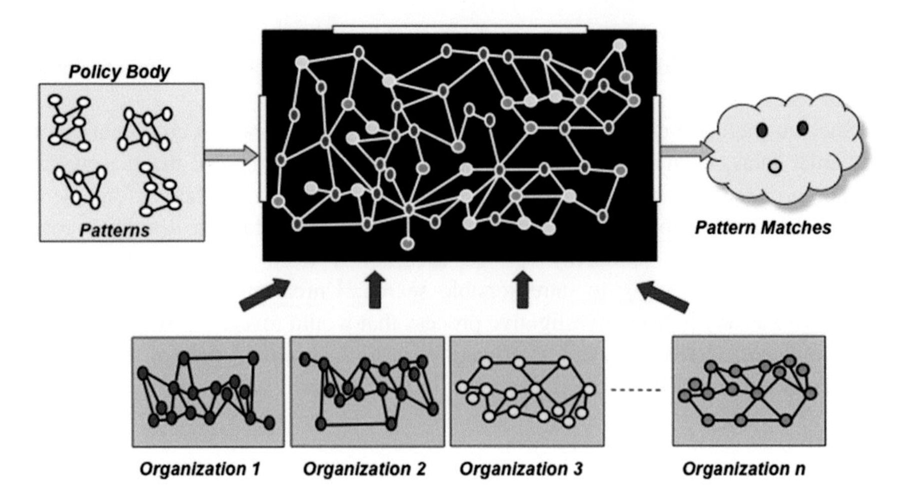

Fig. 1 The privacy assurance "Black Box"

about which a group of reasonable, qualified technical experts will testify that any vulnerabilities inherent in the device's design have been mitigated, using reasonable techniques to assure its security to within a degree of probability asserted as reasonable by a community of such experts.

But what good is a Black Box? Assuming the existence of such a device, it makes possible the ability to "share" private information in unique and powerful ways. However, a new paradigm that governs the notion of analysis and how it can be performed is required.

Figure 1 illustrates the basic privacy assurance concept. At the top center of the diagram is the "Black Box" construct. Across the bottom are representations of independent organizations that span multiple legal and/or jurisdictional boundaries. Each of these organizations via their respective legal charters is authorized to maintain a specific body of information, represented by the colored "dot" networks depicted within each. These information "dots" are connected via "links" that represent relationships that the organization has discerned and maintains, consistent with its legal authorization.

The legal charter of each organization may limit its ability to access or share information and thereby identify the corresponding relationships across the established boundaries. Sharing this information across such a boundary could in fact constitute a breach of law or, alternatively, a breach of public, legislative trust or acceptance. Nevertheless, if such organizations were actually able to share their information, new relationships within the information could be identified from analysis. New patterns of suspicious activity that might impact national security/public safety could be identified and acted upon. This information would constitute "actionable intelligence."

The solution offered here involves placing relevant information from each contributing organization inside of the Black Box. *Information can then be connected and processed within, but only without the possibly of human examination or disclosure.* The internal methods used to do the processing are established in contemporary analytic tradecraft. Techniques such as graph analysis and statistical correlation can discover otherwise hidden relationships among billions of data elements. But if such a Black Box is designed to be "non-queryable" by any means, how then can it be of any value?

To address the utility question, the Black Box also has exactly one additional input (on the left in Fig. 1) and exactly one and only one output (located on the right). At the left interface, patterns of specific interest are input to the box. These patterns are template-like encodings of generic information relationships that a duly authorized policy body has reviewed and approved for submission into the box. Put another way, the patterns are a set of analytical rules that define the Black Box's reasonable search behavior. The only patterns that are admissible to the Black Box are those that the policy body has reviewed and has unanimously confirmed as meeting a certain threshold. In this case, the threshold is the set of observable conditions within the Black Box that meet the legal standard for reasonable suspicion or reasonable concern.

Within the Black Box, in addition to the information that it receives from each contributing organization, and the patterns it receives from the policy body, is an algorithm that continuously observes for conditions that match any of the submitted patterns. Upon detecting such a pattern, the Black Box outputs an identifier for the pattern and a set of identifiers for the information that triggered the pattern's detection. This is a continuous process. It is executed in real-time without human intervention, again leveraging current analytic tradecraft. Upon such a detection event, the Black Box would notify the appropriate contributing organizations of the particular identifiers, but without revealing any of the private information it holds within. These organizations could then investigate further, using their existing analytic capacities and legal authorization structures. If permissible by policy and law, additional information could accompany the output notification to expedite investigation. The specification for such auxiliary output information is incorporated into the original pattern definition, enabling the policy body to review and approve in advance, and ensuring privacy compliance throughout.

Output generated by the Black Box would be available to the policy body or alternatively, to a duly constituted oversight body to continuously verify compliance. In other words, while considerable information is flowing into the Black Box, the only aspect that would ever have external visibility is its reasonable suspicion/concern output. This output would be expressed in terms of identifiers that only have meaning to the submitting organization. In this manner, organizations and the citizenry they serve can receive the benefits or information sharing, but without exposing this information to misuse or the risk of privacy invasion in the process.

Under this paradigm, the only information that can be submitted to the Black Box is information that a participating organization has already been authorized to possess (i.e. this process does not address the sharing and analysis of illegally obtained information). Similarly, the only information that is ever outputted from the Black Box is that which has been deemed *in advance* to constitute reasonable suspicion/concern and to meet the standards of law and public policy for protecting individual privacy.

4 Privacy Certification Levels

This work recognizes that the level of privacy assurance obtainable is directly related to the degree at which privacy device "impenetrability" can be achieved, involving a risk–cost benefit tradeoff. Depending upon the nature of the information to be protected, not all information sharing and analysis applications will require the same degree of rigor to ensure adequate privacy protections. For example, the transmission of personal medical information would presumably have a substantially higher level of privacy concern over, say, sharing of publically available property records. Consequently, a multi-level privacy certification rating is envisioned. Analogous with U.S. cryptographic systems (Committee on National Security Systems 2010), the following four levels of privacy certification are proposed:

- *Type 1 Privacy*: a device or system that is certified for national/international governmental use to securely share and analyze private information consistent with the highest level of protections awarded by law and treaty. Type 1 is used to protect information that would result in exceptionally grave damage if disclosed. Achievement of this rating implies that all components of the end-to-end system have been subjected to strict verification procedures, are protected against tampering and subject to strict supply chain controls with continuous oversight.
- *Type 2 Privacy*: a device or system that is certified for governmental and commercial use to securely share and analyze personal information consistent with high levels of protections in conformance with jurisdictional policies and procedures and commercial law. Type 2 is used to protect information that would result in serious privacy damage if disclosed. Achievement of this rating implies that all interface components of the system have been subjected to strict verification and supply chain controls and that all other components have been subjected to reasonable best industry practices for operation verification and supply chain control and oversight.
- *Type 3 Privacy*: a device or system that is certified for public use to securely share and analyze sensitive information. Type 3 is used to protect information that would result in privacy damage if disclosed. Achievement of this rating implies that all components of the system have been subjected to reasonable best industry practices for operation verification and supply chain control.
- *Type 4 Privacy*: a device or system that is registered for information sharing and analysis, but not certified for privacy protection. No assumptions regarding component verification or supply chain controls are made about systems at this privacy protection level.

At a general level, Type 3 systems are composed of components that are designed and integrated using best industry practice. To achieve a higher assurance rating, best industry practice is not considered adequate. For a Type 2 system, while internal components may be commercial items, all interface components must be subject to a rigorous verification process to ensure the validity of all transactions that cross the Black Box boundary. For a Type 1 system, this same rigor must be applied to the entire system, including the design and implementation of internal components and their procurement supply chain. The primary differentiators between these levels ultimately translate to cost. That is, Type 1 systems will generally be more expensive than Type 2 systems, which in turn will be more costly than Type 3 systems, etc. These cost differences are warranted in order to gain higher assurances of privacy protection due to the varying risks associated with the intended applications at each level.

5 Privacy "Black Box" Design

The generic design of a Black Box is shown in Fig. 2. All information that flows into and/or out of the box must pass through carefully designed interfaces that isolate the Black Box internals from the external environment. External data sources at the left side of Fig. 2 are connected to the box via a set of input isolators. These isolators allow correctly encrypted data to flow into the box only from organizations that are properly authorized and authenticated. These isolators enforce a strict one-way flow of data providing no means of internal access or

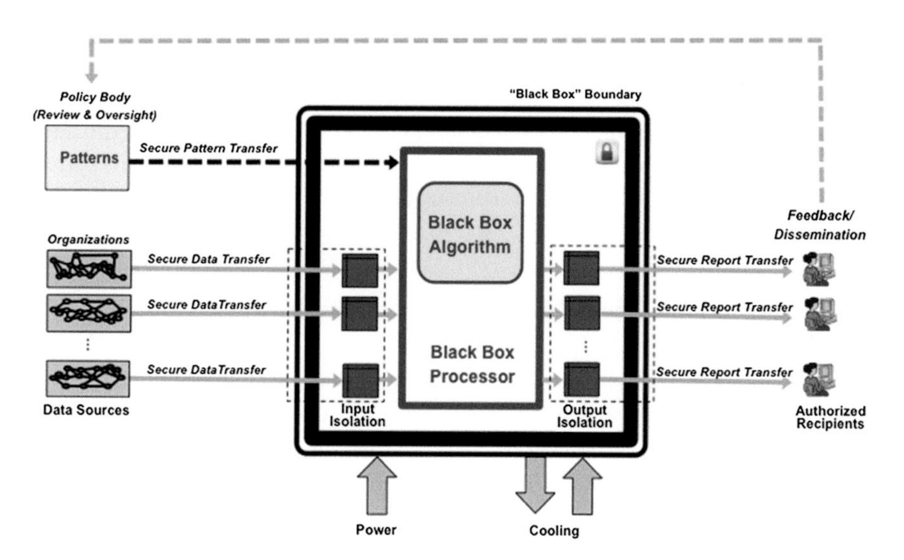

Fig. 2 Generic Black Box design

visibility to any data, status, operating conditions or parameters of the Black Box contents within. While the incoming data received from an organization may well contain private information, accompanying this information for each individual is an identifier (e.g. a unique number sequence) that is assigned by that organization. When a pattern is detected by the Black Box that involves an individual, only the respective identifier for this individual is referenced in the output report. In this manner, no personal information ever leaves the Black Box boundary once it enters.

Within the Black Box is a computer processor that runs an algorithm whose operation is strictly defined by the patterns approved by the policy body. These patterns encode reasonable suspicion/concern policy statements that define the algorithm's behavior. The configuration of this algorithm and its execution of the patterns is carefully controlled and monitored by the policy body to ensure that the Black Box behaves only as they have unanimously specified.

On the right side of Fig. 2 are output isolators that ensure that all output reports that are generated by the internal algorithm flow only to the correct, authorized recipients and that no private information is exposed. The output reports reference individuals via the unique identifiers known and provided by the source organizations. Contained in the reports are indicators of the patterns that the Black Box detected. The policy body controls the specification of these indicators as part of the pattern review and approval process. Unanimous agreement of these indicators is required in advance of the Black Box performing any data analysis. An output feedback loop to the policy body is shown in Fig. 2 for oversight and compliance.

The key aspect of this design is that regardless of what information might flow into the box, the only information that can ever exit is that which was approved and authorized by the policy body as meeting the patterns they have unanimously deemed reasonable. Furthermore, the box itself is implemented in such a manner that these protections cannot be circumvented via tampering. Hence, the implementation cannot provide any back doors, overrides, special authorizations, nor expose any inherent exploitable vulnerabilities, within the limits of the verification techniques and certification process used to specify, design, and engineer its correct operation, commensurate with the assurance level.

6 Example Use Case: Identity Name Resolution

In today's information age, organizations frequently provide overlapping services to individuals. Such overlap can be costly, resulting in unnecessary duplication and expenditure of resources. Resolving this overlap, however, can be extremely complex and time-consuming. Where individuals live, where they work, and how and where they receive these services, and how and when these might change can all greatly vary. Further complicating this process is the incompleteness, errors, and ambiguity in the data that each organization may associate with an individual. The spelling of names, accuracy of birthdates, absence of a consistent universal identifier (e.g. in the U.S., a Social Security Number), etc. all compounds this resolution complexity. Given the sensitive nature of personal information and the complex policies and laws regarding its proper handling, organizations unfortunately are often forced to resort to costly, time-consuming manual methods to identify and resolve discrepancies.

6.1 The Black Box Pilot System

In March of 2015, the first formal application of the Black Box technology was successfully deployed to automate this process in near real-time fashion. The deployment involved three public health organizations working to prevent the spread of HIV within and across their jurisdictional boundaries. Each of these jurisdictions maintains sensitive databases about individuals infected with this disease for their areas. These databases are populated as a result of mandatory reporting procedures followed by the health care providers operating within each of the respective boundaries. To mitigate the spread, it is important that jurisdictions communicate with their neighbors to ensure that individuals remain in care, continuing to receive treatment to help keep their HIV viral counts sufficiently low. As individuals live, work, and receive health care services at varying locations throughout these jurisdictions, resolving identities across the databases has often been a painstakingly slow and difficult process, heightened by the sensitivity of the condition and the importance of protecting each individual's privacy. For this pilot activity, a Type 3 privacy assurance level system was configured. Figure 3 contains an overview of the system's design.

The pilot system consisted of a single, self-contained computer that was physically mounted within a steel reinforced enclosure with multiple security locks (one for each participating jurisdiction). This unit was housed in a non-descript, limited access Tier 3 data center facility managed by Georgetown University with continuous 24/7 video motion detected alarm surveillance. The enclosure was configured such that the computer within could not be removed without resulting in loss of its electrical power. The computer itself was delivered sealed from the factory and was installed and configured only in the presence of security representatives from each organization. The computer was equipped with the most minimal of services, with nearly all external features disabled including the removal of keyboard and mouse input, video display, and unnecessary operating systems components. The disk contents were secured with high-grade encryption. All wireless interfaces (e.g. WiFi and Bluetooth) were disabled, and no external I/O devices were attached nor were ports accessible once secured within the locked enclosure.

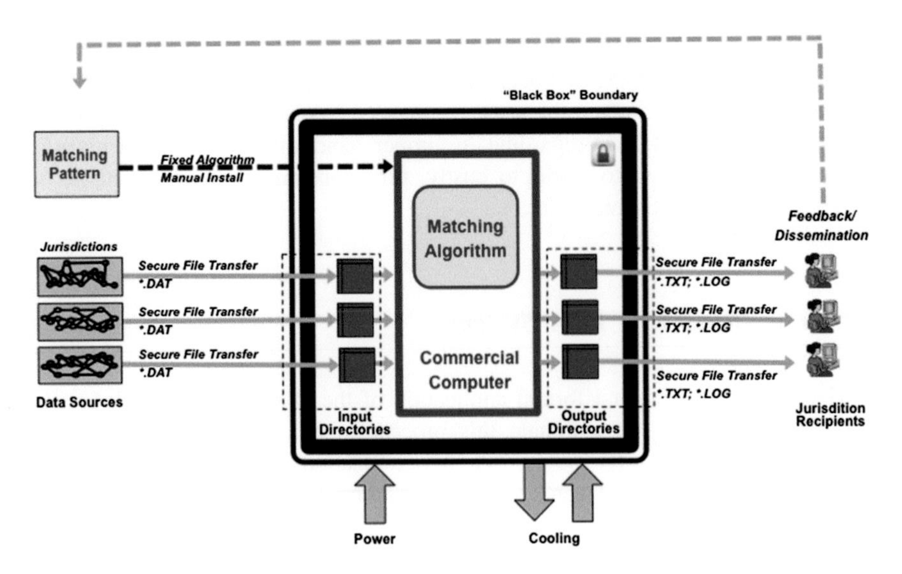

Fig. 3 Pilot (Type 3) Black Box system design

The operating system, network, and supporting firewall infrastructure were configured to allow only secure file transfer access into and out of specific fixed directories, with one directory allocated for each participating organization. The only operations that were permitted by an organization were reading, writing and deleting files in their respective assigned directories. All file accesses were performed via high-grade commercial public-key end-to-end encryption. No other external operations were possible with the enclosure/computer other than the unplugging of its power cord. All administration services, login capabilities, web services, e-mail, etc. had been disabled and/or removed. Specifically, the computer was configured to execute one program and one program only. That is, the computer executed the single identity pattern-matching algorithm that had been review, tested, inspected, and unanimously approved by the policy/oversight body. For this pilot, this body consisted of a representative from each of the participating public health organizations aided by their respective IT staffs

As the reliability of this device and its correctness was of highest concern, the security configuration process was intentionally meticulous and comprehensive requiring the physical presence of an individual from each jurisdiction in order to make changes. Failure to accurately compute results or properly protect the information contained within would have rendered the device useless or even harmful, with significant loss of confidence from each of the participating organizations and their constituency. Operation of this privacy device prototype was intentionally very simple. In order for organizations to identify potential duplication issues, each generated a data file that contained a set of records for the individuals represented in their respective databases. For the initial pilot system, key fields included:

- Last name
- First name
- Date of birth
- Gender
- Ethnicity
- Social Security Number
- Local jurisdiction identifier

Using an agreed upon data file format for these records, each organization securely transferred its file to its respective directory on the privacy device computer. These directories were accessed in a "sally port" like fashion. That is, the organizations placed their data files within these directories, and then upon completion the Black Box algorithm removed the data files from these directories, decrypted the contents and transferred the resulting data into the local memory of the computer. In this manner, there was never any direct communication path between the algorithm and the external environment, as the existence of such paths would have provided a prime target for exploitation by an adversary.

Within the computer, the single program that was run continuously scanned each of the directories for new data files. When a new data file was detected, the file was carefully ingested and in-memory representation of the data is created. The source data file was then immediately securely deleted using multiple file re-writes. The directory scan and file ingest times were specifically engineered so that the sources data files, despite being encrypted, resided on the internal computer disk for a minimal amount of time (e.g. seconds).

In the event a new data file was received from an organization, the old representation was immediately discarded (erased from memory), and the new representation was then compared against the representations held for each of the other organizations. After the comparisons where completed, a report output file was prepared for each organization, identifying only those matches that are made with records of another organization. As they contained no private information, match files remained in the device directories until deleted by the respective organization (or whenever the privacy device system was restarted via power cycling). To further prevent PII exposure, the match files contained only the local organization's unique identifiers and no private source data fields. After a computation cycle, a participating organization was then able to use these identifiers to discuss possible lost-to-care or duplicate-care issues with the other corresponding organizations.

As the Black Box computer intentionally had neither a console nor display and was itself locked in cabinet without any remote monitoring capabilities, ascertaining the operating status of the device could only be performed by the participating jurisdictions. This was possible via a set of log files that was maintained by the algorithm for each jurisdiction. These logs contained the dates and times of ingested data files, when the matching process was performed, and summaries of the degree of matching found. Any errors detected in the input data file formats were reported back to the respective organization through this mechanism. Although operating within a Georgetown University computing facility, no member of the university staff had any ability to examine or monitor the status or contents of the device while it was in operation.

6.2 Pilot System Algorithm

Of all Black Box components, the item perhaps of greatest concern was arguably the algorithm contained within. From a reliability perspective, if this program were to have failed during the pilot's operation (e.g. as a result of an undetected programming error), the jurisdictions (or the developer) would not have had any way of knowing the cause. Although all data transmitted and stored was encrypted, such a failure could have conceivably resulted in a file containing PII persisting far beyond its expected (very short) lifetime upon the device. Such failures, however, could have also severely jeopardized each organization's confidence and trust in the device. If the device was not reliable, organizations would have been justifiably skeptical of its accuracy and its ability to protect such important information. The resulting loss of trust would have rendered the privacy device of little or no value, with the possibility of introducing harm via improper disclosure or wasted time pursuing inaccurate results. Thus, the reliability of the algorithm was of utmost importance throughout the process.

Providing added mechanisms for local real-time status display and remote diagnosis, however, would have increased the complexity of the design and the accompanying risk of compromise, exposing additional penetration paths that an adversary could have potentially exploited. During the development phase of this effort, a system complexity versus system integrity tradeoff became immediately prevalent in the discussion. Adding new features to the design to improve utility or operational use increased overall system complexity. With this added complexity came a tension upon the system's integrity. That is, the consideration of each new feature challenged the assurance of the system's impenetrability level. The pilot activity revealed that this complexity/integrity tradeoff is a fundamental, pervasive issue that must be recognized, addressed, and balanced throughout all phases of any Black Box system's lifecycle. For this pilot, the designers opted to maintain the highest level of simplicity whenever possible to aid the assurance process.

In accordance with its high-reliability and high-integrity design philosophy, the Ada programming language was selected for the algorithm specification and implementation (ISO/IEC 2012). Its unambiguous semantics, extremely strong type and constraint checking, exception protections, formally validated compilers, and overall reliability philosophy were key ingredients leading to this decision. The following is the main subprogram of the pilot system's algorithm:

```
with Black_Box;use Black_Box;

procedure Main is
begin

   Initialize; -- Erase/build directories & logs

   loop
      if Update then-- Check for new data files
         Analyze;          -- Search for matches
         Report;           -- Report matches
         Clear;      -- Clear matches
      end if;
      delay scan_time;
   end loop;

end Main;
```

As can been seen from above, the algorithm was kept very simple and consisted of a single infinite loop. The subprogram `Initialize` was used to create each organization's directory and corresponding log file should, they not already exist. If the directory did exist, its contents were erased, ensuring a fresh start. The package `Black_Box` contained the data structures that represented each organization's data set and the resulting cross organizational matches, along with the algorithm's operations that act upon them (`Update`, `Analyze`, `Report`, and `Clear`). Each of these subprograms was coded so that they would successfully complete, regardless of any internal error or exceptions that might result.

Of all the subprograms, `Update` was perhaps the most worrisome and complex as it involved the ingestion of external data files. While all organizations agreed to a single input format, the algorithm could make no assumptions regarding the input file's compliance as mistakes or errors could otherwise have rendered the system painfully inoperative. Thus when a new data file was detected within the `Update` subprogram, the new input file was very carefully parsed to ensure proper range values and format across all fields. In participating organizations' actual daily practice, it was not uncommon for their source databases to contain blank fields or legacy field formats that contained various wild card characters and special values for missing data elements (e.g. a birth year, but no birth month or day, or "000-00-0000" when a SSN is unknown). The `Update` subprogram's job was to reliably parse through all these various possibilities, reporting format errors back to an organization through its log file, ultimately creating a vector of properly type constrained person records for the corresponding organization. If the process was successful, `Update` returned a **`true`** value, allowing the algorithm to proceed. However, if an unrecoverable problem was detected, **`false`** was returned, preventing the subsequent matching and reporting operations from executing until a new data file was successfully received and processed from the organization.

With a successful (**`true`**) completion of `Update` subprogram, the remaining operations `Analyze` and `Report` were far less perilous as all data structures were now properly type checked and range constrained in comfortable mathematical fashion. The primarily role of the `Analyze` subprogram was to create a vector of records with persons that matched across all represented organizations. Match records contained values that identified the organization, their corresponding person unique identifiers, and a set of values that characterized which and how their fields matched including a score that indicated the likelihood that two individuals were actually the same. Scoring criteria was established via unanimous consent by the participating organizations during the algorithm design process, and then encoded into the `Analyze` subprogram.

The subprogram `Report` had very a predictable role and behavior, predominately creating the matching report output files for each of the organizations within their respective directory. To ensure no memory leaks over time (a common programing flaw), the `Clear` subprogram was used to properly release the dynamic data structures used in the matching process, before the entire process was repeated after a short specified time delay.

6.3 Pilot System Testing and Verification

As a Type 3 device, verification of the prototype system was undertaken using conventional software testing methods, manual code inspection, and comprehensive output file examination commensurate with best software engineering industry practice. Facilitated by participating organizations, a corpus of synthetic test data was used to test the algorithm under many diverse situations. As anticipated, the majority of programming flaws identified in the early testing phase were in the input process dealing with the external data files. However, once data was ingested and represented within the algorithm's strongly typed framework, no errors that would result in catastrophic failure (i.e. program crash or private information exposure) were detected. This was in part a testament to the oversight and involvement of the policy group in specifying and approving the algorithm's behavior. Thorough testing, however, did uncover an obscure programming logic flaw in the matching process due to an incorrect assumption regarding initial variable conditions. While conventional testing methods appeared adequate for a program of such modest size (~ 1000 lines), this process illustrated the critical importance of having a complete formal specification of the algorithm and the use of mathematical assertions and automated program proof-of-correctness techniques necessary to obtain a Type 2 or higher assurance level.

6.4 Pilot System Summary

The Black Box pilot system described here was heralded as a success (Ocampo et al. 2016). In total, the device processed well over 150,000 private information records identifying thousands of previously unknown matches with very high assurance. In total, the computation consumed approximately 20 min, a strong contrast to an otherwise manual process that would have easily extended beyond two years. More importantly, the process was executed entirely without any private information ever being revealed. The pilot exposed and illustrated the diverse spectrum of issues that must be responsibly addressed across a Black Box system's entire lifecycle, from initial design and procurement, to decommissioning and disposal. In summary, the system illustrated that the Black Box technique to private information sharing and analysis is both credible and viable. Moreover, the system successfully challenged the pervading assumption that analysts must have direct access to private, personal information to help further advance national security and public safety objectives. It illustrated that the tension perceived between personal liberty and these objectives need not exist. Rather, it demonstrated that security and safety goals can be met while simultaneously protecting personal information, and that such information need only ever be exposed to select individuals when there exists a very clear legal authority and established need.

7 Privacy Assurance Technology—Type 2

The pilot system discussed provided an illustrative example of an effective Type 3 system design and implementation. Observations throughout its develop process and end-to-end lifecycle helped identify the strengths and limitations of such systems. The technological basis of Type 3 systems is best industry practice. Candidly, as a system is scaled with increasing numbers of individuals and growing data volumes of ever increasing sensitivity, current best industry practice is simply not adequate given the evolving sophistication and insidious nature of contemporary adversaries. Evidence of this assertion can be witnessed each week with yet another major system compromise announced in the news media.

Assessing the pilot's Type 3 design, there are two areas of technical privacy concern. The first involves the method used to transfer private files into the Black Box. Configured using a private data sally port, direct access between the external environment and the internal algorithm is prevented. However, exposing computer file system directories to the outside world presents a potential exploitation path for an adversary, despite whatever firewall, encryption, and user access restrictions that might be imposed. The amount of software involved in a contemporary operating system's file management software and network data transfer applications often comprises many tens of thousands of lines of code (or far greater). Unfortunately, unless this code is specified, designed, and implemented perfectly, an adversary can potentially exploit any weaknesses that may have been overlooked (e.g. buffer overflows, range constraints, undefined states, etc.). As software systems increase in size, catching such mistakes becomes increasingly difficult and expensive. Alas, software "bugs" are indeed commonly found in software systems developed today despite earnest claims of best industry practice.

The second area of concern involves the method for specifying the Black Box algorithm. In the pilot system, while the algorithm was developed outside of box and available for all policy body members review and inspection, it eventually had to be compiled and installed in the Black Box prior to its sealing. This too presents a set of potential exploitation paths, as well as a very real logistical nightmare as the number of participating organizations is increased. Ensuring that the specified algorithm is correct and that the code transferred, installed, and ultimately run on the Black Box involves a large number of technical steps that must be carefully monitored and verified throughout. Unfortunately, this is a very complex process involving many more software modules with potentially hundreds of thousands of lines of code (or even millions). Assurance that this entire ecosystem is without exploitable flaws is far beyond best software engineering practice for any application beyond modest size. Further compounding scalability is the number of parties that would need to be involved to monitor, inspect, and ultimately be present to supervise the loading each time a new algorithm revision is needed becomes very impractical. To address these areas, several modifications are made to the pilot configuration in order to achieve a Type 2 assurance level, as shown in Fig. 4.

At the core of Fig. 4 is what is labeled as a "Secure" computer. As stated previously, perfect security is very elusive. However, the computing industry has made considerable progress developing computers and their companion operating systems for applications where high assurance is vital (e.g. avionics, power systems, medical equipment, etc.). A common framework for specifying computer security and assurance requirements exists and has been widely adopted (ISO/IEC 15408). While there can be no claim that these systems are totally without flaws, the

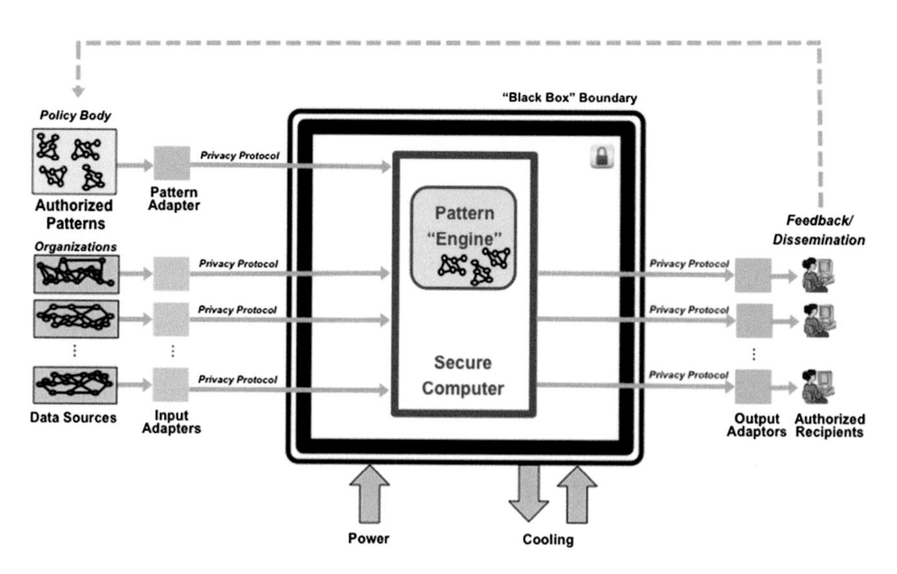

Fig. 4 Type 2 Black Box system design

development process is very rigorous, involving strict quality controls throughout that greatly boosts confidence. Of particular importance is the type and thoroughness of testing of every component and their interactions with others, yielding a certification and accreditation level with supporting documentation evidencing its rigor. While more costly than typical development efforts, this process is warranted by the policy body's assurance demands in order to responsibly mitigate an increased risk level.

Absent from Fig. 4 is the internal directory structure that the pilot system needed to expose to the outside environment. Transfer of bulk data (albeit encrypted) is a potential dangerous activity, as the contents of these data files could potentially contain malware that was injected by adversary. Thus, the data file transfer and file directory structure is replaced with a set of external input and output adapters that interface to the source data organizations and the result recipients, respectively. The primary role of these adapters is to convert source and result data into a set of transaction sequences that flow across the Black Box boundary. These transaction sequences are designed to move single data units, one at a time, verifying the format and validity of each. This is performed using a special privacy transfer protocol,[1] crafted specifically for high-assurance Black Box applications. This protocol is designed to enable all data transactions and related software handling components to be subject to mathematical proof-of-correctness rigor. This is possible with a

[1]The Hypergraph Transport Protocol (HGTP) under development at Georgetown University is specifically designed for this purpose.

complete protocol specification that is formally defined and verified with the inclusion of a vulnerability analysis that spans the full range of possible data values and transaction sequences.

At the right side of Fig. 4, pattern detection reports generated by the pattern engine flow out in a manner similar to the data input process, but in reserve order. That is, triggered pattern identifiers and the associated information identifiers exit the box via the privacy protocol. Once outside the box, this protocol is then converted to a form recognizable by an operator or alternatively to a form that can be processed by the contributing source organizations or investigating bodies that participate in the feedback/dissemination loop for oversight and compliance.

Lastly, rather than expose the internals of the Black Box to a new and potentially incorrect or vulnerable algorithm each time an analytic change is needed, a reusable pattern "engine" is used in Fig. 4 instead. This engine is itself a special algorithm, very carefully engineered to ensure that its pattern-matching operation cannot be modified in any fashion. It is coded one time, test, repaired, and verified perhaps multiple times, but then installed and authenticated in the Black Box once where it remains unchanged until the entire rigorous is repeated to accommodate new features. This process is critical for preventing any type of accidental or adversary-assisted disclosure of private information. Then henceforth, in place of transferring executable code to the Black Box, detection patterns expressed in a special analytic language[2] are instead transmitted, using the same privacy protocol for input data.

Inside the box, the engine interprets remotely specified pattern statements carefully versus trustingly executing them as in the Type 3 design. This interpretation step has the added security benefit that patterns expressed in the specification language cannot cause harm to the Black Box execution, given assurance in advance that the engine is correctly coded. With multiple participating organizations, the engine is configured so that the only patterns it will process are those that are properly expressed in the pattern language with all participating organizations simultaneously agreeing. Unanimous agreement is established by requiring each organization to send the specific pattern that they authorize to the Black Box where they are then compared against all the others. Internally, the engine only proceeds with data analysis and reporting when all of its received patterns are verified and are in proper agreement. Once developed, proven, loaded, and authenticated, the pattern engine algorithm within the Black Box cannot be modified without repeating the entire rigorous, monitored process. However, operational changes to how the Black Box behaves can be accommodated via updates to the pattern specification, considerably reducing the burden associated with refreshing the Black Box's internals.

[2]The ATra language under development at Georgetown University is specifically designed for this purpose.

8 Privacy Assurance Technology—Type 1

High-risk sharing and analysis applications involving extremely private personal information with large volumes of data about large number of individuals will invariably demand the highest level of privacy assurance—Type 1. This would likely include national or international applications that require the greatest level of protections in compliance with law and international treaty. To meet these highest assurances, several additional refinements are needed, as shown in Fig. 5.

At the core of Fig. 5 is now a "trusted" platform. In contrast to the Type 2 secure computer, this platform is a hardware/software device that has been designed and implemented in its entirety with thorough mathematical rigor to ensure its complete proof-of-correctness. As envisioned, this device would be a custom or specially tailored computing system specifically designed for this application. That is, features commonly found in typical off-the-shelf general-purpose computing systems that are not expressly needed to operate the pattern engine would be permanently disabled or removed from the design. Examples of superfluous items might include file storage machinery, all network channels, all input/output channels (excluding only that needed to support the privacy protocol), all display interfaces, and perhaps a large bulk of what is often resident in a typical operating system. In this scenario, the embedded system platform is designed, implemented, and verified precisely for this one privacy application at the greatest level of simplicity to ensure minimal exposure of vulnerability paths.

As software components beyond a few thousands lines of code are typically very hard to prove correct, all direct protocol communication with the Black Box is

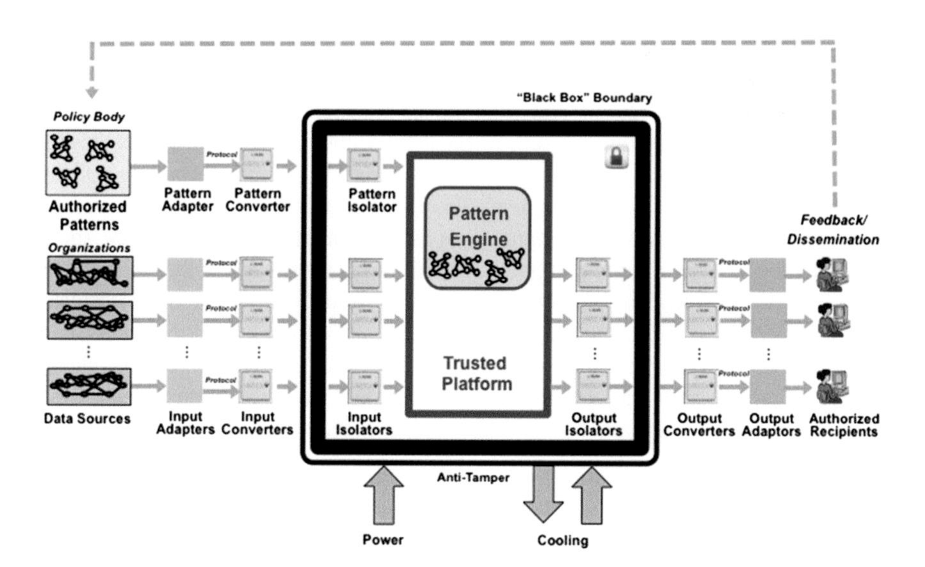

Fig. 5 Type 1 Black Box system design

replaced with a communication channel implemented directly in hardware using combinatorial logic components. This can be achieved with contemporary tradecraft leveraging technology items such as Field Programmable Gate Arrays (FPGA) and Application Specific Integrated Circuits (ASIC). These components have the advantage that their programming can be accomplished using a specification that is more readily subject to mathematical verification. In Fig. 5, each of the source organizations relays their private data to externally located input adapters. These adapters convert the data into a privacy protocol sequence. Each protocol sequence is then sent to an input converter that transforms the transaction into a discrete electrical or optical signal that crosses the Black Box boundary. Within the Black Box, this signal is fed directly to an input isolator that converts the signal into the digital transaction form needed by the pattern engine for processing. The same technique is applied for pattern specification, and the opposite process is used for outputting results to authorized recipients. The proper design and handling of isolators in this manner significantly reduces the vulnerability paths into and out of the Black Box, commensurate with the highest privacy assurance level.

Finally, to further strengthen the actual enclosure that encases all internal Black Box computing components, the use of "anti-tamper" techniques need be employed. This is an area of unique technical tradecraft that prevents adversaries from gaining physical access including special seals and alarm sensors. For fail-safe privacy operation, the primary purpose of these items is the removal of power from the internal components to ensure permanent, irretrievable loss of all Black Box data contents.

9 Operational Considerations

The privacy assurance approach advocated here was derived assuming existing, understood analytic tradecraft and proven, off-the-shelf technology components. While the myriad of technical issues is plentiful with the full spectrum of design and implementation aspects well beyond the scope of this publication, none of the constituent techniques and components described here are particularly new, distinctly novel, or technically unfounded. It is the careful configuration of these components and their unique operationalization within a privacy policy framework that is novel. The proper application, integration, and deployment of these techniques does require both a skilled workforce and a privacy work ethic that is often unfamiliar in everyday computing industry. The pilot activity discussed above helped reveal many of these characteristics. Beyond the technological realm, the primary process issues that must be addressed in tandem include:

- The establishment of a policy body and its associated processes for defining and authorizing patterns that would constitute reasonable suspicion/concern.

- The establishment of an oversight function or oversight body to monitor the operation of a Black Box configuration, including the auditing of input patterns and output reporting to ensure legal compliance.
- The establishment of specific development and deployment process procedures including design, implementation, configuration, physical and cyber protections, testing, and certifications of the Black Box *and its interfaces* to ensure sustained operational system integrity.
- The establishment of operational polices and procedures for identifying, protecting, and mitigating specific vulnerabilities across an end-to-end system deployment.

10 Conclusion

The material in this chapter outlines an approach that enables organizations to share and analyze information in a manner that respects and embraces individual privacy rights. Although discussed here within a privacy policy context, the Black Box assurance approach is applicable to a diverse spectrum of information sharing and analysis challenges including:

- Commercial or community organizations desiring to protect sensitive information across their organizational boundaries to enable cost savings and operating efficiencies while providing improved services.
- Compartmented organizations needing to protect classified or highly sensitive information on a strict need-to-know basis yet work collaboratively towards shared objectives.
- International organizations needing to protect highly sensitive information, perhaps of a treaty, compliance, or deterrence nature, yet work cooperatively to identify areas with common goals and interests.
- Numerous other applications, ranging from health records management and HIPAA compliance to financial information processing for waste, fraud and abuse detection.

With the acceptance and adoption of the Black Box approach to privacy assurance, a new paradigm for private information sharing and analysis is poised to emerge. With this technique, it is possible to declare that all private information about an individual must remain hidden from *all* other individuals unless there is explicit permission for disclosure from the information's owner (e.g. application for a car loan), or legal authority for its examination (e.g. criminal investigation). Organizations that curate private information (e.g. a credit card company) would do so by storing this information in a Black Box container, hidden from view from any of its members. In Black Box fashion, the organization could perform considerable analysis upon this data using the pattern specification and policy body approval machinery without the private data entering an employee's view, except perhaps at initial entry. For many applications, reference to an individual can be accomplished with the organization's assigned identifier, versus repeated exposure of name, address, social security number, telephone number, etc. Computationally, this has the added potential benefit for reducing ambiguity errors where individuals are confused due to similar personal information (e.g. their names).

This construct suggest a further generalization where a set of black boxes may be configured to interact with each other to further reduce the visibility of private information. While this does not suggest that removing human inspection and approval from decision processes is always possible or appropriate, it does offer a mechanism that can greatly reduce private information exposure, limiting access to only those individuals based on confirmed, authoritative need. This new paradigm enables a richness of private information to be shared and analyzed when needed, yet allows carefully restricted user access based on a need-to-know, authorized-to-know basis, and allowed-to-know basis.

Further exploration of these techniques and their many variants offers a unique hope towards addressing the otherwise difficult tension between Security and Liberty. A successful resolution of this tension has profound implications on the modern Information Age.

References

Campos, J. (2014). Civil liberties and national security: The ultimate cybersecurity debate. *Homeland Security Today*, January 27, 2014.

Center for Strategic and International Studies. (2014). *Balancing security and civil liberties—Principles for rebuilding trust in intelligence activities*, Washington D.C., May 15, 2014.

Committee on National Security Systems. (2010). *National information assurance (IA) glossary*. CNSS Instruction No. 4009, April 26, 2010.

Gilmore, J. (2014). Balancing homeland security and civil liberties. *Washington Times*, March 6, 2014.

ISO/IEC 15408. The common criteria for information technology security evaluation

ISO/IEC 8652. (2012). Ada reference manual—Language and standard libraries, 2012(E).

Smart, J. C. (2011). Privacy assurance, international engagement on cyber. Georgetown Journal of International Affairs. http://avesterra.georgetown.edu/tech/privacy_assurance.pdf

Ocampo, J. M., Smart, J. C., et al. (2016). Developing a novel multi-organizational data-sharing method to improve HIV surveillance data for public health action in metropolitan Washington DC. Journal of Medical Internet Research Public Health and Surveillance.

Institutionalizing Ethical Reasoning: Integrating the ASA's Ethical Guidelines for Professional Practice into Course, Program, and Curriculum

Rochelle E. Tractenberg

1 Introduction

"…(E)thics is not a vaccine that can be administered in one dose and have long lasting effects no matter how often, or in what conditions, the subject is exposed to the disease agent". National Academy of Engineering 2009 (p. 34).

"Professional identity formation means becoming aware of … what values and interests shape decision-making." (Trede 2012, p. 163)

"Change begins at the level of individual decisions and behaviors." Heath and Heath (2010, p. 56).

Continuing professional development is an expectation in many fields, including Statistics (American Statistical Association 2011). With the 2014 revision of the statistics undergraduate curriculum report (Horton and The American Statistical Association Undergraduate Guidelines Workgroup 2014,) completed, followed by the 2015 completion of the first revision of the American Statistical Association (ASA) Ethical Guidelines for Statistical Practice since 1999 (http://www.amstat.org/committees/ethics/), it is an ideal time to consider how to introduce these Ethical Guidelines into the initiation—as well as the continuation—of professional development for all those who are trained to engage with data—whether "Big" or 'small'.

Individuals receiving U. S. federal funds for training (e.g., from the National Institutes of Health or National Science Foundation) are required to complete 8 h of face to face training in "the responsible conduct of research" or RCR (Sect. 7009 of the America Creating Opportunities to Meaningfully Promote Excellence in Technology, Education, and Science (COMPETES) Act (42 U.S.C. 1862o–1; https://www.govtrack.us/congress/bills/110/hr2272/text); NIH (2009), http://grants.

R.E. Tractenberg (✉)
Collaborative for Research on Outcomes and -Metrics; and Departments of Neurology, Biostatistics, Bioinformatics & Biomathematics, and Rehabilitation Medicine, Georgetown University Medical Center, 4000 Reservoir Road, Washington, DC, NW 20057, USA
e-mail: rochelle.tractenberg@gmail.com

J. Collmann and S.A. Matei (eds.), *Ethical Reasoning in Big Data*, Computational Social Sciences, DOI 10.1007/978-3-319-28422-4_9

nih.gov/grants/guide/notice-files/NOT-OD-10-019.html). Through/because of these mechanisms, many if not most institutions of higher learning in the United States have created some sort of training program for RCR. Within these RCR training programs, the ethical principles most often invoked are respect for persons; beneficence; and justice—which can be difficult to reconcile with day-to-day statistical practice (although see Steneck 2007). Gelfond et al. (2011) highlighted how failures to follow ASA Ethical Guidelines—as well as violations of other frameworks (including source materials for the NIH RCR training topics list)—have permeated peer-reviewed clinical science over the past 20–30 years. Since the US federal funding agencies have increased their requirements for RCR training over essentially the same period, either the requirement or the "satisfaction" of the requirement (or both) are not having the intended effects. We have argued (Tractenberg and FitzGerald 2012, 2015; Tractenberg et al. 2014; see also Tractenberg 2016a) that the problem arises from the training that is considered to satisfy the requirement (see also National Academy of Engineering 2009; Kalichman 2013).

We have described an alternative training paradigm for promoting the responsible conduct of research (Tractenberg and FitzGerald 2012; see also Tractenberg et al. 2014) which includes an explicit developmental trajectory, rather than a list of topics about which discussions can be facilitated. The trajectory that we advocate focuses on the initiation of the development of expertise, which is recognizable as a transition from more novice-type thinking to more expert-type thinking (Tractenberg et al. 2010; Tractenberg and FitzGerald 2012; Tractenberg et al. 2016). Familiarity with the actual content of the ASA Ethical Guidelines is required for any in-depth application of the constituent principles, but memorizing this content is not a useful or meaningful end for "training in responsible conduct in science". The Presidential Commission for the Study of Bioethical issues, whose May 2014 report (regarding ethics and neuroscience) emphasizes that "…integrating ethics and neuroscience throughout the research endeavor….offers a means by which researchers can recognize and respond to ethical issues that arise throughout the research process. (Presidential Commission for the Study of Bioethical Issues, May 2014, p. 5). Opinion is converging from disparate fields like engineering and neuroscience: **preparing scientists to *reflect* will promote ethical decision-making** (in research and in practice).

This includes statistics; and just as the ASA Ethical Guidelines state, "(t)he principles expressed here should guide both those whose primary occupation is statistics and those in all other disciplines who use statistical methods in their professional work" (p. 1) (see also, Gelfond et al. 2011). As we have described elsewhere (Tractenberg et al. 2014; Tractenberg and FitzGerald 2015), a semester course can be structured around initiation, growth, and development of the reasoning skills that are the essential foundation for the application and eventual internalization of ethical principles in scientific work generally (see e.g., Tractenberg et al. in review).

The idea that "…(t)he entire community of scientists and engineers benefits from diverse, ongoing options to engage in conversations about the ethical dimensions of research and (practice)," (Kalichman 2013: 13) is clearly aligned with an emphasis on ethical reflection—and is <u>inconsistent</u> with a culture where the *same* "*RCR training*" is required for new students, senior faculty, and everyone in between—irrespective of career stage, level of responsibility, role in research, or disciplinary speciality. Within a culture where a one-time ethics training "vaccine" is the standard, neither Kalichman's ideal—nor that of the Presidential Commission on Bioethics—can ever be realized. Instead, if *inculcation*, rather than compliance, became a driving force for ethics education, trainees/students could learn both how to engage in these important conversations about the ethical dimensions of research and practice—and that such conversations *are* important. Preparing scientists to engage competently in these conversations requires purposeful, widespread, and developmental training that can come from, and support, a culture of ethical research and statistical—as well as scientific—practice.

2 Integration into Existing Courses

One of the main barriers to having required ethics training for 100 % of students at 100 % of universities, whether there are degree programs in statistics or biostatistics or not, is that either a new, additional, course is required—with many sections to accommodate all students, or some single general course that can be completed online is needed. A challenge for either of these approaches is that teaching and learning about ethics, or professional practice, is qualitatively different—particularly in terms of assessment—than it is around statistics and biostatistics/data analysis. One reason why memorization of rules or principles is often the means for assessment of learning is that this is the simplest and most consistent method for testing and demonstrating knowledge.

A second barrier to more comprehensive engagement with ethics across institutions and students might be considered to comprise the most common content: it can sometimes seem that "ethics training" is only required for those who work directly with human subjects, or that it is most or only important for those who violate norms for ethical practice (or the law). The most egregious violations of these norms are falsifying data, committing fraud, or plagiarizing (also known as "FFP") in scientific research. Most faculty, if they have received training in ethics or in the "responsible conduct of research", have received instruction in how to avoid—sometimes how to recognize, and prevent in their students—FFP.

Examination of the ASA Ethical Guidelines for Professional Practice, however, highlights the wide variety of behaviors and decisions that constitute "ethical" behavior. As the preamble to the Guidelines articulates, there are many different purposes for the creation and maintenance of these Guidelines; there are also multiple purposes of their integration into courses, programs, and curricula. These include:

A. Encouraging ethical conduct in (throughout) the practice of science, by pointing out how everyone on a research team has their specific role with its attendant obligations and priorities.

B. Promoting professionalism for all of the research team members, including analysts irrespective of their level of training in statistics.
C. Promoting the consideration, prior to the start of analyses, of the analyses and the qualifications of the analyst to plan, execute, and interpret them.
D. Engaging with principles of professional practice for statisticians, which can promote both appreciation for the statistician as a collaborating research team member and understanding how this team member is accountable and responsible for their work. Even if students are not going to be the 'designated statistician' on a project, understanding the role and responsibilities of this team member can strengthen the sense of responsibility and accountability of each member of a research team.

While they are written for the ASA membership, the Guidelines exist to promote a sense of responsibility and stewardship for science in general. We have shown (Tractenberg et al. 2014; Tractenberg 2013) how the eight principles of the Guidelines map neatly with the nine topic areas that the National Institutes of Health (NIH) have outlined as important to cover in "responsible conduct of research" training (NIH 2009; http://grants.nih.gov/grants/guide/notice-files/NOT-OD-10-019.html). As such, integrating the Guidelines into an existing course can promote the achievement of the RCR training requirements for <u>all</u> students at an institution, as well as those receiving federal funding for training. Since data analysis is becoming such an important area in all disciplines (scientific, business, and other), the ASA Guidelines can serve to introduce all students to critical concepts of responsible data analysis, interpretation, and reporting.

The first requirement for integrating the ASA Ethical Guidelines into an existing course is to include the Guidelines (see Appendix) in the syllabus. From the instructor's perspective, this integration—rather than simply adding the document onto the end of the syllabus—is an opportunity to consider the Guidelines themselves and where/when during the course the Guidelines can be brought into the discussion. Table 1 presents the principles and some of their key elements comprising the 2015 ASA Ethical Guidelines for Professional Practice together with discussion questions/prompts that can be used in any training context. Some of the prompts are specific for use with homework problems, but if undergraduate courses in statistics or experimental design include time for working through homework problems or examples within the lectures, any of these discussion prompts can be utilized there as well. After practice with the discussion questions that involves formative feedback, small group discussions would be supported, with students reporting their group's discussion and consensus or result.

In addition to including discussion in undergraduate classes around the Guidelines, principles and specific elements, the prompts in Table 1 can also be integrated into journal club discussions around any article by focusing on the analyses that are reported, and whether or not a member of the authorship team (or those identified in an acknowledgment) is a member of a statistics or biostatistics department.

Engaging students in active discussion around the Guidelines is only one element of their integration into the course. As the Guidelines state, their purpose is to

Table 1 Discussion around the guidelines, the principles, and their elements

Guidelines text	Comment/recommended discussion
Purpose of the Guidelines	
The discipline of statistics links the capacity to observe with the ability to learn and make decisions. It provides a foundation for building a more informed society. There are many ways to be misled by statistics; because society depends on informed judgments supported by statistical methods, all practitioners of statistics, whatever their training and occupation or job title, have an obligation to perform their work in a professional, competent, and ethical manner	***Reflection on the nature of ethical scientific conduct is motivated by the preamble***
The American Statistical Association's Ethical Guidelines for Statistical Practice are intended to help statistics practitioners make and communicate decisions ethically, and to inform employers of statisticians and those relying on statistical analysis about the standards that they should expect	*Do these Guidelines inform you as a: (A) consumer of statistics? (B) producer of statistics? If so, how? If not, why not?*
Application of these ethical guidelines generally requires good judgment and common sense. In some cases, prioritizing Guideline principles may result in a degree of conflict between different principles; the application of these Guidelines can also depend on issues of law and shared values. **Ethical professional practice in statistics requires following these Guidelines to the extent possible**	*List some decisions to which this text might refer.* • *Identify which principles apply for each of those decisions.* • *Identify at least two pairs of potentially conflicting principles.* • *Discuss management and resolution of these two conflicts*
Stakeholders in data analysis, *including the statistician*, all have professional and personal priorities; as such, these may conflict at any of the stages in an analysis	*List some conflicting professional vs. personal priorities (e.g., advance career vs. refuse unethical analysis request)*
All stakeholders have an obligation to act in good faith, and these Guidelines are intended to promote the accountability of data analysts throughout their involvement in any project. Personal integrity is essential for practitioners of statistics, and in their practice integrity is fundamentally based on transparency of assumptions, reproducibility of results, and validity of interpretations	*Explain this obligation from the perspectives of consumer of data analysis and that of producer of analysis*
The principles expressed here should guide both those whose primary occupation is statistics and those in all other disciplines who use statistical methods in their professional work. Therefore, throughout these Guidelines, the term "statistician" shall be read to include all practitioners of statistics and quantitative sciences, regardless of job title or field of degree, comprising statisticians at all levels of the profession *and*	*Explain how these Guidelines can provide their intended guidance for you as both a consumer and as a producer of statistical analysis, considering your present level of training and preparation*

(continued)

Table 1 (continued)

Guidelines text	Comment/recommended discussion
members of other professions who utilize and report statistical analyses and their implications	
A. Professional Integrity & Accountability	
The ethical statistician uses methodology and data that are relevant and appropriate, and uses them objectively, without bias, and in a manner intended to produce valid, interpretable, and reproducible results	
The ethical statistician shall:	
1. Maintain, and be prepared to document for peers or employers, their professional-level competency in relevant methodologies. Ensure he/she possesses adequate statistical and subject-matter expertise before undertaking any analysis	*Describe yourself with respect to this item. Are you qualified to do what you are asked to do in a homework problem or in the laboratory? If so, can you document that qualification?* *(Case studies: document qualifications of individual(s) asked to carry out analyses)*
2. Identify and mitigate any biases on the part of the investigators or data providers that might predetermine or influence the analyses/results	*Describe biases that investigators or those providing the data (to you or in the case study) might have. How might you identify bias, or determine if it is present?*
7. Accept full responsibility for his/her professional performance. Provide only such expert testimony, written work, and oral presentations as he/she would be willing to have peer reviewed	*Considering your professional preparation, explain how this principle is: A) in conflict with one other Guideline principle or with a personal priority; or B) applicable in a case study*
B. Responsibilities to Science/Public/Funder/Client	
The ethical statistician is supportive of valid inferences, and good science in general, keeping the public, funder, client or customer interests in mind (as well as our professional colleagues, patients/the public; and the scientific community)	
The ethical statistician shall:	
1. Clearly state his/her statistical qualifications and experience relevant to all work	*Describe yourself with respect to this item. Are you qualified to do what you are asked to do in a homework problem or in the laboratory? What kind and extent of relevant experience represents "qualification"?* *(Case studies: document qualifications of individual(s) asked to carry out analyses)*

(continued)

Table 1 (continued)

Guidelines text	Comment/recommended discussion
3. To the extent possible, present a client or employer with choices among valid alternative statistical approaches that may vary in scope, cost, or precision	*Consider "to the extent possible". The analyst's responsibility is to present valid alternatives; what might limit the analysts' ability to present these to the client or employer?*
5. Apply statistical sampling and analysis procedures scientifically, without predetermining the outcome	*Describe how you might NOT do this. Can you give two examples of how you could predetermine or bias the outcome?*
C. Integrity of data and methods	
The ethical statistician shall be open and candid about any known or suspected limitations, defects, or biases in the data that may impact the integrity or reliability of the statistical analysis. Objective and valid interpretation of the results requires that data analysis recognizes and acknowledges the degree of reliability and integrity of the data	
The ethical statistician shall:	
1. Report statistical and substantive assumptions made in the execution and interpretation of any analysis. Report the limitations of statistical inference of the study and possible sources of error	*Case study, journal article, or homework/consulting problem: List the assumptions involved in the analysis or problem*
2. Clearly and fully report the steps taken to guard integrity of data and validity of results	*Identify two or more steps that the statistician can take to guard data integrity; list two steps to take to ensure validity in the results*
3. When reporting analyses of volunteer data or other data not representative of a defined population, include appropriate disclaimers	*List two disclaimers that would be "appropriate" in a report of volunteer or other less-representative data*
4. Promptly correct any errors discovered after publication or producing the final report. As appropriate, disseminate the correction publicly or to others relying on the results	*Describe how you would do this, or how you have done it in the past*
D. Responsibilities to Research Subjects	
The ethical statistician shall protect and respect of human and animal subjects at all stages of the statistical process. This includes respondents to the census or to surveys, and data taken from administrative records, as well as subjects of physically or psychologically invasive research	
The ethical statistician shall:	

(continued)

Table 1 (continued)

Guidelines text	Comment/recommended discussion
1. Know about and adhere to applicable rules and approvals for the protection of human subjects and applicable animal welfare guidelines	*Is this Guideline applicable to the analyst of ANY data, and/or by analysts with ANY level of training? Identify sources of such information*
2. Avoid the use of excessive or inadequate numbers of research subjects, and excessive risk to research subjects (in terms of health, welfare, privacy, and ownership of their own data), by making informed recommendations for study size	*Explain the relevance of this element for Principle D*
5. Before participating in a study involving human beings or organizations, analyzing data from such a study, and while reviewing manuscripts for publication or internal use, consider whether appropriate research subject approvals were obtained. The statistician should consider the treatment of research subjects (e.g., confidentiality agreements, expectations of privacy, notification, and consent, etc.) in contemplating the appropriateness of the data source(s)	*Describe how you would (and/or do) "consider whether appropriate research subject approvals were obtained". Is this relevant only for data from human subjects?*
6. In contemplating whether to participate in an analysis of a particular data source, refuse to do so if participating in the analysis could reasonably be interpreted by individuals who provided information as sanctioning a violation of their human rights	*Why is this element included in Principle D? What is its relevance for ethical professional practice of statistics or science?*
E. Responsibilities to Research Team Colleagues	
The ethical statistician shall:	
1. Maintain, and be prepared to document for peers or employers, their professional-level competency in the Ethical Guidelines for Statistical Practice	*Describe how you would do this, or how you have done it in the past*
4. Avoid compromising statistical or scientific validity for expediency. To the extent possible, promote the most effective and efficient use, of all statistics, by the entire research team	*Describe how a compromise in validity could happen to promote expediency. How could it (a request, demand, or just pressure to compromise) be avoided?*
F. Responsibilities to Other Statisticians or Statistics Practitioners	
The ethical statistician shall:	
4. Respect differences of opinion	*How might differences of opinion occur during research or statistical analysis?*

(continued)

Table 1 (continued)

Guidelines text	Comment/recommended discussion
5. Instill in students and non-statisticians an appreciation for the practical value of the statistical concepts and methods they are learning or using	*Why is this element included in Principle F?*
G. Responsibilities Regarding Allegations of Misconduct	*Why do the Guidelines for Professional Practice include this item?*
The ethical statistician shall:	
1. Avoid condoning or appearing to condone incompetent or unethical practices in statistical analysis.	*How does following this element demonstrate ethical professional practice?*
2. Avoid and act to discourage all types of professional and scientific misconduct	*Discuss how failure in any of the Ethical Guidelines for Professional Conduct might constitute misconduct*
4. Be aware of definitions of, and procedures relating to, misconduct. If involved in a misconduct investigation, follow prescribed procedures	*Why is this important to professional practice?*
H. Responsibilities of Employers, Including Organizations, Individuals, Attorneys, or Other Clients Employing Statistical Practitioners	*Discuss these responsibilities from the perspective of someone who asks you—or someone whom you ask—to carry out an analysis. Case studies or journal club articles: consider these from the perspectives of the first author and from the principle analyst on the paper*
Those employing any person who analyzes data are expected to:	
1. Recognize that the results of valid statistical studies cannot be guaranteed to conform to the expectations or desires of those commissioning the study or the statistical practitioner(s)	*How important is this element from the employer's perspective? Is it more or less important than for the analyst?*
2. Recognize that valid findings result from competent work in a moral environment. Employers, funders, or those who commission statistical analysis have an obligation to rely on qualified statisticians for any data analysis. These obligations may be especially relevant in analyses that are known or suspected to have tangible physical, financial, or psychological impact(s)	*How do less qualified statisticians become more qualified? Is coursework sufficient?*
3. Recognize that these Guidelines exist, and were instituted, for the protection and support of the statistician and the consumer alike	*Discuss how these Guidelines protect or support the statistician or the consumer*

"help statistics practitioners make *and communicate* decisions" (emphasis added). Integrating the same discussion questions around other problems (e.g., on homework), and requiring—and evaluating—written reflections in response to the prompts, will help prepare students to engage more fully and more thoughtfully in the in-class discussion provided that consistent formative feedback is also provided. We have created a rubric (Tractenberg and FitzGerald 2015) that describes the levels of written work that the *mentors* (instructors) should have achieved in ethical reasoning around the eight ASA Ethical Guidelines principles (A–H) in Table 2 of that paper. In two other manuscripts (Tractenberg 2013; Tractenberg et al. 2014) we have included syllabi and/or rubrics that describe the elements that should be included in narrative work that beginners compose.

Instructors can develop their own rubrics for both guiding student writing in response to these prompts around the Guidelines, by adapting existing writing rubrics that may be in use at their institutions or previously published (e.g., Timmerman et al. 2011), or by creating their own in consultation of excellent resources like "Introduction to Rubrics" (Stevens and Levi 2005). It is critical that students at all levels are given the rubrics early, and are also given opportunities to revise their writing with formative, constructive, input. As mentioned, this is qualitatively different than the sort of assessment that is typical for statistics and biostatistics courses. Most universities have education excellence or teaching and learning centers that can support faculty initiatives to integrate new writing (narrative/reflective) assignments—an their evaluation—into courses where they have not been used. Our preliminary data (Tractenberg et al. in review) support the idea that preparing researchers to reflect on their reasoning will promote ethical decision-making and research, and also that teaching ethical reasoning, rather than exposure to the main "topics of RCR", **can** lead to Kalichman's ideal of "… ongoing options to engage in conversations about the ethical dimensions of research and (practice)," by supporting sustainable learning, i.e., when the one course is done, the learning and practice continue (Knapper 2006; Schwänke 2009).

3 Integration into Programs

Table 1 presents discussion prompts that, as discussed, can be integrated into courses where statistics/analysis problems are worked and/or discussed—including homework or example problems as well as articles in journal club settings. The integration of the ASA Ethical Guidelines for Professional Practice into a single existing course is feasible for introducing undergraduate and graduate non-majors to the Guidelines within the single "introduction to statistics" course they might be required to complete. However, the Guidelines could easily be integrated into multiple courses (e.g., a multi-course sequence) or into a program simply by replicating the discussions, and the writing assignments, in each of those courses. In this context, a program is defined as a series of courses that are not necessarily leading to a degree (which is defined as a curriculum, see next section).

All principles in the Guidelines, but not every element of each principle, are relevant for undergraduates in *and out* of the statistics/biostatistics majors, and for graduate students outside of the discipline. This is the reason for isolating specific elements within each principle in Table 1 for discussion and consideration. As they engage with consultation and with actual practice, graduate students in statistics and biostatistics would find the remainder of the elements in each Guideline principle becoming relevant; and discussions around those elements can be structured within multi-course sequences for these students along the same lines as the discussion prompts given in Table 1. Since the prompts for discussion can be used with virtually any type of homework problem or worked example, the repetition of their integration across multiple courses helps to increase both exposure to the Guidelines and depth of processing with the elements and their applicability throughout the scientific enterprise.

The developmental trajectory that Tractenberg and FitzGerald (2012) outlined for ethical reasoning, which is also the basis for the semester course syllabus included in Tractenberg et al. (2014) and Tractenberg and FitzGerald (2015), can be used to support growth, rather than just repetition, in the reflection that is targeted in the written responses to the suggested prompts in Table 1. If programmatic integration of the Guidelines is proposed to incorporate this development in ethical reasoning skills in addition to familiarity with the Guidelines themselves and how they must be applied and prioritized in different situations, a single course dedicated to the ethical reasoning skills together with an introduction to the Guidelines themselves can be helpful. The published course syllabi (Tractenberg 2013; Tractenberg et al. 2014) can be used to augment existing programs by supporting the development of stand-alone courses in ethical reasoning with the ASA Ethical Guidelines for Professional Practice that are also consistent with the NIH (and NSF) requirements for training in RCR. Tractenberg (2013) outlines the alignment of the ASA Guideline Principles with NIH topic areas, articulating how an ASA Guidelines-based course would also meet these federal RCR training requirements if they pertain (see Tractenberg 2016a). Then, the suggestions from the previous section would be applicable to each of the successive courses within the program. It is important to both A. ensure that sufficient time and instruction, together with practice and feedback, is dedicated to the introduction of the paradigm (ongoing, integrated emphasis on professional practice, the Guidelines, and/or ethical reasoning); and B. purposefully and consistently integrating additional opportunities for considering and discussing the Guidelines throughout the other courses in the program, particularly promoting growth and the demonstration of that growth by the students.

The Mastery Rubric for Ethical Reasoning (MR-ER, Tractenberg and FitzGerald 2012) outlines a career-spanning training trajectory of development in ethical reasoning. We conceptualized "ethics education" as a set of six learnable, improvable types of knowledge, skills or abilities (KSAs): Prerequisite knowledge; recognizing an ethical issue; identification of decision-making frameworks; identification and evaluation of alternative actions; making and justifying decisions; and reflecting on the decision (Santa Clara University (no date); see also Kligyte et al.

2008b; Hollander and Arenberg 2009 for similar lists of ethical reasoning elements). The list focuses on decision-making and reasoning—found by Antes et al. (2010) to be conspicuously absent or to worsen after traditional "RCR training" (see also Mumford et al. 2009; Antes et al. 2009; Schmaling and Blume 2009). The dynamic trajectory can apply to faculty (preparing them to guide learners towards ethical research and practice in their specific domain) as well as students (orienting them explicitly to what ethical research and practice look like in the domain, see Tractenberg, this volume for more on modeling professional habits of mind). Because ethical reasoning is a skillset not tied to topical material, this trajectory can be applied in any field. The developmental trajectory of a Mastery Rubric is *used by the instructor* to align assignments with objectives, and to assess student work—but it is also *used by the student* to assess their own skills and, possibly, their need for additional training, practice, or opportunities to refine or demonstrate their KSAs (Tractenberg et al. 2010). Assignment-specific rubrics can be adapted, adopted (where existing and relevant for reflective writing, e.g., http://oregonstate.edu/ctl/reflective-writing-rubric; see also Timmerman et al. 2011), or created, to support the creation of opportunities for teaching and learning that are specific to the Guidelines and professional identity development.

4 Integration into Curricula

In the previous section, suggestions for integrating the ASA Ethical Guidelines for Professional Practice into a "program", or series of courses that are not necessarily leading to a degree, were discussed. The Mastery Rubric (MR, Tractenberg et al. 2010) is a curriculum development and evaluation tool. Wolf (2007, p. 17) articulated three processes in curriculum development: "visioning"; alignment, coordination and development of objectives; and the actual development of the curriculum. The MR approach to curricular design is grounded on the alignment of learning goals—articulated up front for stages of student development through the curriculum—with assessment, both opportunities and types (see Boud and Falchikov 2006). The development (e.g., Tractenberg et al. 2010) or revision (Tractenberg et al. 2016) of a curriculum using the Mastery Rubric also requires that stakeholders identify and align the instructional and learning objectives with the elements of assessment validity outlined by Messick (1994):

1. What is/are the knowledge, skills, and abilities (KSAs) that students should possess (at the end of the curriculum)?
2. What actions/behaviours by the students will reveal these KSAs?
3. What tasks will elicit these specific actions or behaviours?

This way, "success" can be characterized, not in terms of completing a series of courses, but in terms of developing the habits of mind and the base of knowledge that can continue to foster excellence in the domain of interest. That is, the intention is that all students will be firmly within the "proficient" column on all of the skills, and that claims of their proficiency will be supported with concrete evidence (Mislevy 2003). When students have moved to the 'proficient' side of the rubric, the curriculum can be evaluated in explicit terms—providing concrete characterizations of each student based on work products from the courses, rather than subjective ratings or other variable (or sample dependent) methods.

With a list of topics or of training opportunities, the "proficiency" level is *inferred* by the number of items checked off (individual) or included on (institution) that list. Self-monitoring in this context is limited to "what activity/topic haven't I done (individual) or offered (institution) yet?" By contrast, the Mastery Rubric for Ethical Reasoning (Tractenberg and FitzGerald 2012) lists target KSAs with performance levels ranging from novice to journeyman and master-level: performance for each KSA at each level can be used by students to show how their proficiency is increasing. Instructors can use the increasing expertise in responses to both structure increasingly advanced engagement with ethical reasoning or the Guidelines principles, and also to specifically support growth in the KSAs, and to evaluate student work as it is produced across courses within a program or curriculum. With an initial course to introduce both the Guidelines and the ethical reasoning knowledge, skills and abilities, the development of a portfolio—similar to that required for the ASA's Professional Statistician (Pstat®) accreditation—that documents growth and development of these skills can also be utilized. The summative assessment of a curriculum might be a research paper or capstone project, and a portfolio that is assembled to capture growth and development, as recognized by the student—using their own work and its maturation over the curriculum as evidence of growth—can augment the capstone project.

With a Mastery Rubric approach to curriculum development, throughout the completion of coursework, student self-monitoring can focus on, "how well do I do/know this KSA, what do I need to do to become more proficient?" (for the individual) and, "what opportunities for improving this KSA do we need to offer to permit individuals to show they are becoming even more proficient at this KSA?" (for the instructor or institution). A Mastery Rubric for Ethical Reasoning can be utilized to augment an existing curriculum in statistics or biostatistics for graduate students by creating one new course and integrating the Guidelines throughout the other courses in the curriculum. As noted, this approach would ensure that the ASA Ethical Guidelines are represented throughout the curriculum and that the curriculum includes ample opportunities for instruction, practice with feedback, and the demonstration of growth in abilities to use these guidelines in future professional practice, consistent with best practices in higher education (e.g., McKeachie and Svinicki 2011; see also Wiggins and McTighe 1998).

5 Discussion

Recent work in the development of professional identity has suggested that "(S)tudents could learn more from their experiences if they were more explicitly guided to look out for certain aspects of professionalism and given further opportunities to discuss and critique their observations and experiences. " (Grace and Trede 2011, p. 12). Trede and McEwen (2012) also argue that the development of professional identity is supported by instruction and practice in justifying professional decisions: "…students need to learn to articulate the reasons behind their actions." (Trede and McEwen 2012, p. 10; see also Trede 2012). Undergraduate and graduate degree programs may be challenged to add a new course, and may be unable to replace existing courses, with one on "ethics". The ASA Ethical Guidelines, however, relate to professionalism and professional practice—which entail ethical behavior, but are essential in initiating and continuing professional—and professional identity—development to ultimately strengthen the field. Continuing professional development is well-known to be difficult to foster, monitor, and document in medical and nursing education (Macy Foundation 2008, 2010; see also Novossiolova and Sture 2012).

This article emphasizes active learning (e.g., Fink 2013, p. 116) through ongoing discussion of the application(s) of the Guidelines, rather than memorizing their content. Although the Guidelines are strongly recommended to be included in the syllabus of a single course, or across syllabi in programs and curricula, the engagement with the Guidelines, their purpose, principles, and constituent elements as structured through discussion prompts in Table 1 represent a constructivist approach to learning (Knowles and Holton 2005; pp. 191–192).

The integration of ethical reasoning principles, and encouragement that students monitor their own progress will advance *all* teaching and learning objectives within a program or curriculum through the development and emphasis of metacognition (Ambrose et al. 2010, pp. 190–216). Even if the metacognitive growth is initiated with activities around the Guidelines, our preliminary research suggests that this will transfer, or be sustained, by doctoral students (Tractenberg et al. in review—we have no data or experience with undergraduate or masters level students). Specifically, an emphasis on metacognition and reflection, rather than on remembering and recognition, in teaching and learning about ethics in statistical professional practice are recommended because they are consistent with adult learning theory (e.g., Knowles and Holton 2005) and with the initiation of the development of expertise (Ericsson 2003). In particular, the ability to recognize and seek out remediation for one's own skills or knowledge gaps, supporting "deliberate practice" (Ericsson 2003), requires metacognitive skills.

Ambrose et al. (2010), among others, argue that metacognition is one of seven principles that promote effective learning; the National Research Council (2001) states that "(m)etacognition is crucial to effective thinking and competent performance." (p. 78). In his review of his work studying expertise, Ericsson (2003) notes that, "(t)he key challenge for aspiring expert performers is to avoid the arrested development associated with automaticity… and instead acquire cognitive skills to support continued learning and improvement. …These mechanisms are designed to increase the experts' ability to monitor and control these processes." (p. 113) the elements most supportive of "effective" ethics education (May and Luth 2013)—not just factual "ethics" knowledge but also, reasoning in the face of uncertainty (see Mumford et al. 2008) and a sense of purpose for engaging in self-directed learning (see Paterson et al. 2002; Ambrose et al. 2010, pp. 190–216). Dunlosky et al. (2013) reported that "self-explanation" (metacognition) is among the best-supported learning techniques they explored.

The objective of this paper is to support the integration of principles of professional and accountable conduct of statistical analyses throughout an educational program—in and outside of the major, since not all future statistical analysts are trained formally in statistics. For undergraduate, graduate, and some post graduate students outside of statistics and biostatistics programs, there might only be a single required course in statistics or in ethics/responsible conduct of research. Like other "required" courses, this might be the only formal exposure students will have to the topic—even if statistics or data analysis will be an important part of their later careers. The recommendations made here for integration of the ASA Ethical Guidelines for Professional Practice into existing courses, multi-course sequences or programs, and curricula involve discussion activities that promote engagement with the Guidelines, and reflection on both their application and their applicability. Moreover, these recommendations are consistent with a recent meta-analysis of characteristics of effective ethics training: Antes et al. (2009) reported that the inclusion of material that is specific to how individuals might actually deal with ethical problems when they are encountered might improve the efficacy of that training. While the integration of the Guidelines as described here can only indirectly support ethical research and practice, a widespread effort to integrate active and ongoing discussions about responsibility and accountability in research may support a change from the current state where just 35 % of programs require "at least some ethics training for at least some students" (Lee et al. 2015) to a much higher level of both engagement and permeation. It can also promote a culture in which ethical research and practice are deliberate learning objectives.

Acknowledgements RET was supported by an NSF EESE grant ("A multidimensional and dynamic ethics education paradigm spanning the science career"; award #1237590). The author is Vice-chair of the ASA Committee on Professional Ethics and chaired the working group on revising the ASA Ethical Guidelines for Professional Practice (2014–2015). There are no other actual or potential conflicts for the author.

Appendix A. ASA Ethical Guidelines for Statistical Practice (*current as of 11 November 2015. Please check* http://www.amstat.org/about/ethicalguidelines.cfm *for the most recent version*)

Proposed Revised

Ethical Guidelines for Statistical Practice

Prepared by the Committee on Professional Ethics of the American Statistical Association,

Distributed to ASA Membership August 2015

Purpose of the Guidelines

The American Statistical Association's Ethical Guidelines for Statistical Practice are intended to help statistics practitioners make and communicate decisions ethically as well as to inform those relying on statistical analysis, including employers, colleagues, and the public, of the standards that they should expect, thereby promoting accountability. The discipline of statistics links the capacity to observe with the ability to learn and make decisions, providing a foundation for building a more informed society. Because society depends on informed judgments supported by statistical methods, all practitioners of statistics, whatever their training and occupation or job title, have an obligation to perform their work in a professional, competent, and ethical manner. All practitioners of statistics should avoid and act to discourage any type of professional and scientific misconduct.

In some situations, Guideline principles may conflict, requiring individuals to prioritize principles according to context. However, in all cases, stakeholders have an obligation to act in good faith, to act in a manner that is consistent with these Guidelines, and to encourage others to do the same. Good statistical practice is fundamentally based on transparency of assumptions, reproducibility of results, and validity of interpretations. Above all, professionalism in statistical practice presumes the goal of advancing knowledge while avoiding harm; using statistics in pursuit of unethical ends is inherently unethical.

The principles expressed here should guide both those whose primary occupation is statistics and those in all other disciplines who use statistical methods in their professional work. Therefore, throughout these Guidelines, the term "statistician" shall be read to include all practitioners of statistics and quantitative sciences, regardless of job title or field of degree, comprising statisticians at all levels of the profession *and* members of other professions who utilize and report statistical analyses and their implications.

Professional Integrity and Accountability

The ethical statistician uses methodology and data that are relevant and appropriate, and uses them without favoritism or prejudice, and in a manner intended to produce valid, interpretable, and reproducible results. The ethical statistician does not knowingly take on work for which he/she is not sufficiently qualified, is honest with the client in any limitation of expertise, and consults other statisticians when necessary or in doubt.

The ethical statistician shall:

1. Identify and mitigate any preferences on the part of the investigators or data providers that might predetermine or influence the analyses/results.
2. Employ selection or sampling methods and analytic approaches appropriate and valid for the specific question to be addressed, so that results extend beyond the sample to a population relevant to the objectives with minimal error under reasonable assumptions.

3. Respect and acknowledge the contributions and intellectual property of others.
4. When establishing authorship order for posters, papers and other scholarship, strive to make clear the basis for this order, if determined on grounds other than intellectual contribution.
5. Disclose conflicts of interest, financial and otherwise, and manage or resolve them according to established (institutional/regional/local) rules and laws.
6. Accept full responsibility for his/her professional performance. Provide only such expert testimony, written work, and oral presentations as he/she would be willing to have peer reviewed.

Integrity of Data and Methods

The ethical statistician shall be open and candid about any known or suspected limitations, defects, or biases in the data that may impact the integrity or reliability of the statistical analysis. Objective and valid interpretation of the results requires that data analysis recognizes and acknowledges the degree of reliability and integrity of the data.

The ethical statistician shall:

1. Acknowledge statistical and substantive assumptions made in the execution and interpretation of any analysis. When reporting on the validity of data used, acknowledge data editing procedures, including any imputation and missing data mechanisms.
2. Report the limitations of statistical inference and possible sources of error.
3. In publications, reports, or testimony, identify who is responsible for the statistical work if it would not otherwise be apparent.
4. Report the sources and assessed adequacy of the data; and account for all data considered in a study and explain the sample(s) actually used.
5. Clearly and fully report the steps taken to guard integrity of data and validity of results.
6. Where appropriate, address potential confounding variables not included in the study.
7. In publications or testimony, identify the ultimate financial sponsor of the study, the stated purpose, and the intended use of the study results.
8. When reporting analyses of volunteer data or other data that may not be representative of a defined population, include appropriate disclaimers and, if used, appropriate weighting.
9. To aid peer review and replication, share the data used in the analyses whenever possible/allowable, and exercise due caution to protect proprietary and confidential data, including all data that might inappropriately reveal respondent identities.

10. Strive to promptly correct any errors discovered after publication or producing the final report. As appropriate, disseminate the correction publicly or to others relying on the results.

Responsibilities to Science/Public/Funder/Client

The ethical statistician is supportive of valid inferences, transparency, and good science in general, keeping the public, funder, client or customer interests in mind (as well as professional colleagues, patients, the public; and the scientific community).

The ethical statistician shall:

1. To the extent possible, present a client or employer with choices among valid alternative statistical approaches that may vary in scope, cost, or precision.
2. Strive to explain any expected adverse consequences of failure to follow through on an agreed-upon sampling or analytic plan.
3. Apply statistical sampling and analysis procedures scientifically, without predetermining the outcome.
4. Strive to make new statistical knowledge widely available to provide benefits to society at large and beyond his/her own scope of applications.
5. Understand and conform to confidentiality requirements of data and any restrictions on its use established by the data provider (to the extent legally required), and protect use and disclosure of data accordingly. Guard privileged information of the employer, client, or funder.

Responsibilities to Research Subjects

The ethical statistician shall protect and respect the rights and interests of human and animal subjects at all stages of their involvement in a project. This includes respondents to the census or to surveys, those whose data are contained in administrative records, as well as subjects of physically or psychologically invasive research.

The ethical statistician shall:

1. Keep informed about and adhere to applicable rules, approvals, and guidelines for the protection and welfare of human and animal subjects.
2. Strive to avoid the use of excessive or inadequate numbers of research subjects, and excessive risk to research subjects (in terms of health, welfare, privacy, and ownership of their own data), by making informed recommendations for study size.

3. Protect the privacy and confidentiality of research subjects and data concerning them, whether obtained directly from the subjects, other persons, or existing records. Anticipate and solicit approval for secondary and indirect uses of the data when obtaining approvals from research subjects; and obtain approvals appropriate to allow for peer review and independent replication of analyses.
4. Be aware of legal limitations on privacy and confidentiality assurances. Do not over-promise or assume legal privacy and confidentiality protections where they may not apply.
5. Consider whether appropriate research subject approvals were obtained before participating in a study involving human beings or organizations, analyzing data from such a study, and while reviewing manuscripts for publication or internal use. The statistician should consider the treatment of research subjects (e.g., confidentiality agreements, expectations of privacy, notification, and consent, etc.) in contemplating the appropriateness of the data source(s).
6. In contemplating whether to participate in an analysis of a particular data source, refuse to do so if participating in the analysis could reasonably be interpreted by individuals who provided information as sanctioning a violation of their rights.

Responsibilities to Research Team Colleagues

The ethical statistician shall:

1. Recognize that other professions have standards and obligations, that research practices and standards can differ across disciplines, and that statisticians do not have obligations to follow standards of other professions that conflict with these Guidelines.
2. Ensure that all discussion and reporting of statistical design and analysis is consistent with these Guidelines.
3. Avoid compromising scientific validity for expediency.
4. Strive to promote transparency in design, execution, and reporting or presenting of all analyses.

Responsibilities to Other Statisticians or Statistics Practitioners

The practice of statistics requires consideration of the entire range of possible explanations for observed phenomena, and distinct observers drawing on their own unique sets of experiences can arrive at different and potentially diverging judgments about the plausibility of different explanations. Even in adversarial settings, discourse tends to be most successful when statisticians treat one another with mutual respect and focus on scientific principles, methodology and the substance of data interpretations. Out of respect for fellow statistical practitioners, the ethical statistician shall:

1. Promote sharing of data and methods as much as possible and as appropriate without compromising propriety. Make documentation suitable for replicate analyses, metadata studies, and other research by qualified investigators.
2. Be willing to help strengthen the work of others through appropriate peer review; in peer review, respect differences of opinion and assess methods, not individuals. Strive to complete review assignments thoroughly, thoughtfully, and promptly.
3. Strive to instill in students and non-statisticians an appreciation for the practical value of the concepts and methods they are learning or using.
4. Use professional qualifications and contributions as the basis for decisions regarding statistical practitioners' hiring, firing, promotion, work assignments, publications and presentations, candidacy for offices and awards, funding or approval of research, and other professional matters.
5. Avoid harassment and discrimination.

Responsibilities Regarding Allegations of Misconduct

The ethical statistician shall:

1. Avoid condoning or appearing to condone incompetent or unethical practices in statistical analysis.
2. Recognize that differences of opinion and honest error do not constitute misconduct; they warrant discussion, but not accusation.
3. Be aware of definitions of, and procedures relating to, misconduct. If involved in a misconduct investigation, follow prescribed procedures.
4. Maintain confidentiality during an investigation, but disclose the investigation results honestly once they are available.
5. Following a misconduct investigation, support the appropriate efforts of all involved, including those reporting the possible scientific error or misconduct, to resume their careers in as normal a manner as possible.
6. Avoid, and act to discourage, retaliation against or damage to the employability of those who responsibly call attention to possible scientific error or misconduct.

Responsibilities of Employers, Including Organizations, Individuals, Attorneys, or Other Clients Employing Statistical Practitioners

Those employing any person to analyze data are expected to:

1. Recognize that these Guidelines exist, and were instituted, for the protection and support of the statistician and the consumer alike.
2. Recognize that valid findings result from competent work in a moral environment. Employers, funders, or those who commission statistical analysis have an obligation to rely on qualified statisticians for any data analysis. This obligation may be especially relevant in analyses that are known or suspected to have tangible physical, financial, or psychological impact(s).
3. Recognize that the results of valid statistical studies cannot be guaranteed to conform to the expectations or desires of those commissioning the study or the statistical practitioner(s).
4. Recognize that it is contrary to these Guidelines to report or follow only those results that conform to expectations without explicitly acknowledging competing findings and the basis for choices regarding which results to report, use, and/or cite.
5. Recognize that the inclusion of statistical practitioners as authors, or acknowledgement of their contributions to projects or publications, requires their explicit permission because it implies endorsement of the work.
6. Support sound statistical analysis and expose incompetent or corrupt statistical practice.
7. Strive to protect the professional freedom and responsibility of statistical practitioners who comply with these Guidelines.

Appendix B. Brief Outline of Courses

Semester course (gets students from Novice to Beginner):

Using a published semester course syllabus (Tractenberg et al. 2014), a semester course to introduce the framework for ethical reasoning can be structured as a series of at least 10 meetings (30 h). These 10 meetings need not be completed in a semester but can be monthly and be completed over a year (however in this model, the attention that the students and faculty can bring to each assignment might wane or be less than if it is completed in a semester). This course will achieve several objectives, including: introduce ethical reasoning as a construct, and its constituent knowledge, skills and abilities (KSAs); give practice and feedback in each KSA for each meeting; orient individuals to the Mastery Rubric and its performance level descriptors; introduce the concept of self-assessment/reflection and how to use and assess evidence to this end; and familiarize participants with their professional code (s) of conduct. During class meetings through the semester, cases must be identified that require at least some of the Code or Guideline principal areas. The KSA of the week is also described, practiced, and explicitly incorporated into the ethical reasoning (ER) framework during class discussion. In each meeting, the cases (1–2) are discussed by having the students read and expound on their respective 500-word case analyses. Prior to the start of each class meeting, the case analyses, which are assigned the week before, must be formatively evaluated and returned for revision. The work that was turned in would normally be what the student discusses in class, and the extent that the instructor's feedback has been processed (or received) can also be part of the discussion. The students, based on the formative feedback and the in-class discussion, have 1–2 days to revise their essays; this revised version is turned in and is then summatively evaluated. A final 1000-word essay directs students to utilize their prior essays as evidence of earlier and later performance (demonstrating either growth or stability in performance) within a single portfolio. The essay outlines which assignments exemplify the novice, and which the beginner, level performance of each KSA. Irrespective of where the student feels they perform, the purpose of this portfolio is to articulate the argument and present the evidence that supports a claim of "being at the Beginner stage". Not all KSAs must/will be performed at the Beginner stage, some portfolios might contain some or mostly novice level performance. A portfolio should include reflection on both how gains in sophistication/performance were achieved as well as whether/how stability (lack of advancement) happened during the semester where it was the target to improve on these skills.

Post-course follow-up (gets students from Beginner to Journeyman):

After completing the semester course described above—and/or, for those whose portfolios or 1000-word essays outline that and how they have reached the Beginner level on all KSAs, then a second series of 10 potentially less-structured sessions (not necessarily whole class meetings) can be offered to students to potentially bring an individual from Beginner up to Journeyman level performance on each KSA. These meetings will vary in length based on how many students attend and the amount of writing the students have done/turned in for formative input. After these meetings, a portfolio that captures changes in performance since the end of the initial course is created. This portfolio, unlike the one demonstrating and documenting change from Novice to Beginner—which is all focused on the evidence from the course and so utilizes the ten case analyses that were generated, should instead utilize experiences and materials from throughout the individual's entire life as a learner. Journeyman-level portfolio assembly requires at least about 26 h, and this time is ALL active on the participant's part (true for students and faculty who create Journeyman portfolios).

References

Ambrose, S. A., Bridges, M. W., DiPietro, M., Lovett, M. C., & Norman, M. K. (2010). *How learning works: Seven research-based principles for smart teaching*. San Francisco, CA: Wiley.

American Statistical Association. (2011). *Guidelines for voluntary professional accreditation by the American Statistical Association*. (Rev 1). http://www.amstat.org/accreditation/pdfs/Guidelines_for_ASAVoluntary_Professional_Accreditation.pdf. Accessed July 2012.

Antes, A. L., Murphy, S. T., Waples, E. P., Mumford, M. D., Brown, R. P., Connelly, S., & Devenport, L. D. (2009). A meta-analysis of ethics instruction effectiveness in the sciences. *Ethics and Behavior, 19*(5), 379–402.

Antes, A. L., Wang, X., Mumford, M. D., Brown, R. P., Connelly, S., & Devenport, L. D. (2010). Evaluating the effects that existing instruction on responsible conduct of research has on ethical decision making. *Academic Medicine, 85*, 519–526.

Boud, D., & Falchikov, N. (2006). Aligning assessment with long-term learning. *Assessment and Evaluation in Higher Education, 31*(4), 399–413.

Boud, D., & Falchikov, N. (Eds.). (2007). *Rethinking assessment in higher education: Learning for the longer term*. New York, NY: Routledge.

Dunlosky, J., Rawson, K. A., Marsh, E. J., Nathan, M. J., & Willingham, D. T. (2013). Improving students' learning with effective learning techniques: Promising directions from cognitive and educational psychology. *Psychological Science in the Public Interest, 14*, 4–58. doi:10.1177/1529100612453266. Retrieved from 30 May 2014 http://psi.sagepub.com/content/14/1/4.full.pdf+html.

Ericsson, K. A. (2003). The search for general abilities and basic capacities: Theoretical implications from the modifiability and complexity of mechanisms mediating expert performance. In R. J. Sternberg, & E. L. Grigorenko (Eds.), *The psychology of abilities, competencies and expertise* (pp. 93–125). Cambridge, UK: Cambridge University Press.

Gelfond, J. A. L., Heitman, E., Pollock, B. H., & Klugman, C. M. (2011). Principles for the ethical analysis of clinical and translational research. *Statistics in Medicine, 30*, 2785–2792.

Grace, S., & Trede, F. (2011). Developing professionalism in physiotherapy and dietetics students in professional entry courses. *Studies in Higher Education*. doi:10.1080/03075079.2011.603410.

Heath, C., & Heath, D. (2010). *Switch: How to change things when change is hard*. New York, NY: Broadway Books.

Hollander, R., Arenberg, C. R. (Eds.). (2009). *Ethics education and scientific and engineering research*. Washington: National Academy of Engineering.

Horton, N., & The American Statistical Association Undergraduate Guidelines Workgroup. (2014). Curriculum guidelines for undergraduate programs in statistical science. Downloaded on 14 Feb 2015 from http://www.amstat.org/education/pdfs/guidelines2014-11-15.pdf.

Kalichman, M. (2013). Why teach research ethics? In *National academy of engineering (Eds). Practical guidance on science and engineering ethics education for instructors and administrators* (pp. 5–16). Washington, DC: National Academies Press.

Kligyte, V., Marcy, R. T., Waples, E. P., Sevier, S. T., Godfrey, E. S., Mumford, M. D., & Hougen, D. F. (2008-b). Application of a sensemaking approach to ethics training in the physical sciences and engineering. *Science and Engineering Ethics, 14*(2), 251–78.

Knapper, C. (2006, August). *Lifelong learning means effective and sustainable learning: Reasons, ideas, concrete measures*. Seminar presented at 25th International course on Vocational Training and Education in Agriculture, Ontario, Canada. Downloaded on 2 October 2013 from http://www.ciea.ch/documents/s06_ref_knapper_e.pdf.

Knowles, M. S., & Holton, E. F, I. I. I. (2005). *The adult learner (6E)*. Burlington, MA: Elsevier.

Lee, L. M., McCarty, F. A., & Zhang, T. R. (2015). Ethical numbers: Training in US graduate statistics programs, 2013–2014. *The American Statistician, 69*(1), 11–16. doi:10.1080/00031305.2014.997891.

Macy Foundation. (2008). Continuing education in the health professions: Improving healthcare through lifelong learning [Monograph]. Conference sponsored by the Josiah Macy, Jr. Foundation, Southhampton, Bermuda. 2007, November. Retrieved on 21 October 2014 from http://www.macyfoundation.org/docs/macy_pubs/pub_ContEd_inHealthProf.pdf.

Macy Foundation. (2010). *Lifelong learning in medicine and nursing*. Final conference report. Association of American Medical Colleges, American Association of Colleges in Nursing. Funded by the Josiah Macy, Jr. Foundation. Retrieved on 21 October 2014 from http://www.aacn.nche.edu/education/pdf/MacyReport.pdf.

May, D. R., & Luth, M. T. (2013). The effectiveness of ethics education: A quasi-experimental field study. *Science and Engineering Ethics 19*(2), 545–568.

McKeachie, W. J., & Svinicki, M. (2011). *McKeachie's Teaching Tips, 12E*. Boston, MA: Houghton Mifflin.

Messick, S. (1994). The interplay of evidence and consequences in the validation of performance assessments. *Educational Researcher, 23*(2), 13–23.

Mislevy, R. J. (2003). Substance and structure in assessment arguments. *Law, Probability, and Risk, 2*, 237–258.

Mumford, M. D., Connelly, S., Brown, R. P., Murphy, S. T., Hill, J. H., Antes, A. L., Waples E. P., & Devenport, L. D. (2008). A sensemaking approach to ethics training for scientists: Effects on ethical decision-making. *Ethics and Behavior 18*, 315–339.

Mumford, M. D., Connelly, S., Murphy, S. T., Devenport, L. D., Antes, A. L., Brown, R. P., Hill, J. H., & Waples, E. P. (2009). Field and experience influences on ethical decision-making in the sciences. *Ethics and Behavior 19*(4), 263–289.

National Academy of Engineering (NAE) (2009). *Ethics education and scientific and engineering research*. Washington, DC: National Academies Press.

National Institutes of Health. (2009). *Update on the requirement for instruction in the responsible conduct of research.* NOT-OD-10-019. http://grants1.nih.gov/grants/guide/notice-files/NOT-OD-10-019.html. Accessed 25 January 2012.

National Research Council. (2001). *Knowing what students know*. Washington, DC: National Academy Press.

Novossiolova, T., & Sture, J. (2012). Towards the responsible conduct of scientific research: is ethics education enough? *Medicine Conflict and Survival, 28*(1), 73–84.

Paterson, M., Higgs, J., Wilcox, S., & Villeneuve, M. (2002). Clinical reasoning and self-directed learning: Key dimensions in professional education and professional socialisation. *Focus on Health Professional Education: A Multi-Disciplinary Journal, 4*(2), 5–21.

Santa Clara University. (no date). Ethical reasoning. Downloaded from http://www.scu.edu/ethics/. 29 November 2009.

Schwänke, U. (2009). *Sustainable learning—how storyline can support it*. Paper presented at the Nordic Storyline Conference, Gothenburg, NE (pp 1–2). Downloaded at http://www.storyline-methode.de/mediapool/43/436167/data/Sustainable_learning_-_nachhaltiges_Lernen.pdf. 10 October 2013.

Schmaling, K. B., & Blume, A. W. (2009). Ethics instruction increases graduate students' responsible conduct of research knowledge but not moral reasoning. *Accountability in Research, 16*, 268–283.

Steneck, N. H. (2007). *ORI introduction to the responsible conduct of research, Revised*. Accessed from https://ori.hhs.gov/sites/default/files/rcrintro.pdf. February 2010.

Stevens, D. D., & Levi, A. J. (2005). *Introduction to Rubrics: An assessment tool to save grading time, convey effective feedback and promote student learning*. Portland, OR: Stylus Publishing.

Timmerman, B. E. C., Strickland, D. C., Johnson, R. L., & Payne, J. R. (2011). Development of a 'universal' rubric for assessing undergraduates' scientific reasoning skills using scientific writing. *Assessment and Evaluation in Higher Educatino, 36*(5), 509–547.

Tractenberg, R. E. (2013). Ethical reasoning for quantitative scientists: A mastery rubric for developmental trajectories, professional identity, and portfolios that document both. In *Proceedings of the 2013 joint statistical meetings*, Montreal, Quebec, Canada.
Tractenberg, R. E. (2016a). Creating a culture of ethics in Biomedical Big Data: Adapting 'guidelines for professional practice' to promote ethical use and research practice. In L. Floridi & B. Mittelstadt (Eds.), *Ethics of biomedical big data*. London: Springer.
Tractenberg, R. E. (2016b). Integrating ethical reasoning into preparation for participation to work in/with Big Data through the Stewardship model. In J. Collmann & S. Matei (Eds.), *Ethical reasoning in big data*. New York: Springer.
Tractenberg, R. E., & FitzGerald, K. T. (2012). A Mastery Rubric for the design and evaluation of an institutional curriculum in the responsible conduct of research. *Assessment and Evaluation in Higher Education, 37*(7–8), 1003–1021.
Tractenberg, R. E., & FitzGerald, K. T. (2015). *Responsibility in the conduct of quantitative sciences: Preparing future practitioners and certifying professionals*. Presented at 2014 Joint Statistical Meetings, Boston, MA. Proceedings of the 2015 Joint Statistical Meetings, Seattle, WA.
Tractenberg, R. E., FitzGerald, K. T., & Collmann, J. (in review). Evidence of sustainable learning with the Mastery Rubric for ethical Reasoning.
Tractenberg, R. E., Gushta, M. M., & Weinfeld, J. (2016). The mastery rubric for evidence-based medicine: Institutional validation via multi-dimensional scaling. *Teaching and Learning in Medicine*.
Tractenberg, R. E., McCarter, R. J., & Umans, J. (2010). A Mastery Rubric for clinical research training: guiding curriculum design, admissions, and development of course objectives. *Assessment and Evaluation in Higher Education, 35*(1), 15–32. doi:10.1080/02602930802474169.
Tractenberg, R. E., Russell, A., Morgan, G., FitzGerald, K. T., Collmann, J., Vinsel, L., et al. (2014) Amplifying the reach and resonance of ethical codes of conduct through ethical reasoning: Preparation of big data users for professional practice. *Science and Engineering Ethics*. http://link.springer.com/article/10.1007%2Fs11948-014-9613-1.
Trede, F. (2012). Role of work-integrated learning in developing professionalism and professional identity. *Asia-Pacific Journal of Cooperative Education, 13*(3), 159–167.
Trede, F., & McEwen, C. (2012). Developing a critical professional identity: Engaging self in practice. In J. Higgs, R. Barnett, S. Billett, M. Hutchings & F. Trede (Eds.), *Practice-based education* (pp. 27–40). Rotterdam, The Netherlands: Sense Publishers.
Wiggins, G., & McTighe, J. (1998). What is backward design? In *Understanding by design*. Upper Saddle River, NJ: Merrill Prentice Hall. (pp. 7–19). Retrieved from http://nhlrc.ucla.edu/events/startalkworkshop/readings/backward-design.pdf.
Wolf, P. (2007). A model for facilitating curriculum development in higher education: A faculty-driven, data-informed, and educational developer–supported approach. *New Directions for Teaching and Learning Winter, 2007*(112), 15–20.

Data Management Plans, Institutional Review Boards, and the Ethical Management of Big Data About Human Subjects

Jeff Collmann, Kevin T. FitzGerald, Samantha Wu, Joel Kupersmith and Sorin Adam Matei

1 Introduction

The National Science Foundation (2012) defines big data as "large, diverse, complex, longitudinal, and/or distributed data sets generated from instruments, sensors, Internet transactions, email, video, click streams, and/or all other digital sources available today and in the future" and, more generally, as referring to the "three Vs": volume, velocity, and variety of data (Steinmann et al. 2015). Examining big data from an ethical perspective has identified additional characteristics that tend to emerge as large quantities of data get used for various purposes, characteristics we have labeled the "4Rs"; that is, reuse, repurposing, recombination and reanalysis. In any given situation, any subset of these characteristics may occur singly or together. As a set, however, the four R's explain why privacy protection overlaps with, but exists distinct from information protection. The "4Rs" also raise the possibility that the ethical provenance and ethical horizon of specific big data sets transcend the narrow circumstances of their original collection. Given the significance of the Belmont Report and human subjects protection for science conducted under the Common Rule, the 4Rs pose questions about the long-term obligations of scientists working with big data about human affairs that may not apply to either governmental or commercial organizations.

In this chapter we will explore these ideas through a sequence of steps. First we will explain the 4Rs and the underlying concepts of ethical provenance and ethical horizon. Second we will present a set of decision trees for helping investigators prepare Data Management Plans when writing proposals for federal research grants using big

J. Collmann (✉) · K.T. FitzGerald · S. Wu · J. Kupersmith
Georgetown University, Washington DC, USA
e-mail: collmanj@georgetown.edu

S.A. Matei
Purdue University, West Lafayette, Indiana, USA

J. Collmann and S.A. Matei (eds.), *Ethical Reasoning in Big Data*,
Computational Social Sciences, DOI 10.1007/978-3-319-28422-4_10

data research on human subjects, particularly for the National Science Foundation. Third, we will demonstrate how to use the DMP decision tree by exploring several cases of big data human subject research that illustrate the 4Rs. In these case analyses we will also explore the implications of our entire analysis for the Notice of Proposed Rule-making for the Common Rule and, with that, for the work of Institutional Review Boards (IRBs) in Common Rule organizations. We make the general point that human subjects research with big data poses new and important challenges that require investigators to work with both their IRBs and computer security experts at their institutions when planning and executing their research projects.

2 The 4Rs

Big data archives get constructed in many ways. In some cases such as genomic databases the scale, volume and complexity of a single archive warrants calling it "big." In other cases, an archive becomes big through the integration of multiple components that investigators compile in a single archive or otherwise link for coordinated analysis. The data elements may be homogenous or highly heterogeneous as when linking unstructured with structured data in a single research project. Investigators may deploy these data elements in four ways that our analysis suggests have ethical implications, including reusing, repurposing, recombining or reanalyzing.

Reuse: Reuse refers to taking data originally collected for a specific scientific purpose and using them again for comparable purposes in comparable domains. The reuse activities may engage either the original investigators or other investigators. The possibility of reuse, particularly of data originally acquired in scientific activities covered by the Common Rule, raises the question of the responsibilities that investigators have for what happens to data once they become available to secondary investigators. Additionally, it poses the question of the responsibilities of secondary investigators for complying with, or reaffirming the conditions of a data set's original collection.

Repurposing: In contrast to reuse, repurposing refers to taking data originally collected for a specific purpose in a specific domain and analyzing them for unrelated purposes in a domain other than their domain of origin. In addition to the questions posed by reusing data, repurposing big data poses questions about the legitimacy of analyzing data acquired under one privacy context and employing it in a different privacy context, particularly Common Rule research.

Recombining: The term big data frequently evokes a process of combining and recombining data from various sources to achieve greater analytic yield. In addition to the questions posed by reusing and repurposing data, recombining data poses questions about the possibility of developing new information not available to the investigator simply from the constituent data sets. From a privacy perspective, recombining data potentially enables re-identification of individuals from data that contains no specific identifiers or has been intentionally stripped of identifiers. Indeed, a research project may posit such re-identification as an explicit goal as in attempting to track persons with an infectious disease across time and space. The possibility of re-identification through recombining data raises questions about privacy protection distinct from information protection.

Reanalysis: Big data archives have been assembled, particularly in public health and healthcare, with comparative and longitudinal purposes in mind. Although investigators may identify some specific objectives at the time of the archives' creation, they also expect and hope that new uses may emerge as scientific knowledge grows, lines of inquiry develop and techniques for extracting new information from collected data sources become more sophisticated.

3 Mirrored Concepts: Ethical Horizon and Ethical Provenance

For investigators operating under the Common Rule, the 4Rs imply a general ethical responsibility for data stewardship that we identify with two mirrored inter-related concepts, ethical horizon and ethical provenance. Both concepts refer to the responsibility of investigators to ensure ethical management and privacy protection for big data across the data lifecycle of collection, compilation, analysis and application. They differ in the vantage point and perspective of the investigator on the unfolding movement of data over time and projects. Ethical horizon refers to the perspective of investigators at the time of data creation who look to future uses of the data, their own and all subsequent investigators. Ethical provenance refers to the perspective of an investigator who looks to the past from which data comes. As a project unfolds or data become created, used and used again, these vantage points shift. Naming both perspectives speaks to the likelihood that investigators conducting research on human subjects entailing big data will occupy both vantage points and adopt both perspectives sometimes within the very same project as they create, reuse, repurpose, recombine and reanalyze data. Naming both perspectives, however, also suggests the need for institutional frameworks that embrace the ethical horizons and ethical provenances of big data in collection, compilation, analysis and application both within and between collaborating organizations.

The concepts of ethical horizon and ethical provenance entail at least five important implications, particularly in light of the societal debate about privacy in government and commercial uses of big data.

1. In contrast to the President's Council of Advisors on Science and Technology (PCAST) discussion of commercial "Notice and Consent" practices that urges refocusing policy, procedures and practices on the phase of data use, scientists must concern themselves about privacy and ethical evaluation of big data across

all phases of big data collection, compilation, analysis and application (President's Council of Advisors on Science and Technology 2015).

2. In concert with discussion of PCAST's discussion of commercial "Notice and Consent", however, scientists do bear responsibility for protecting human subject's privacy and the various consequences of privacy breaches. When a human subject gives informed consent to use personal information, they do not and cannot assume responsibility for its protection and rightfully expect scientists to protect it across the data life cycle.
3. When drawing data from multiple sources originally built under various circumstances, scientists working under the Common Rule must evaluate the ethical status of each original collection regime as well as their own projected analysis and use of the data.
4. Scientists working under the Common Rule must conduct privacy and ethical analysis of data as an ongoing feature of research with big data because of its dynamic, unpredictable and uncertain outcomes in scientific inquiry, including the possibility that data may change its character from less to more sensitive over the course of an analysis.
5. Because scientists working under the Common Rule generally fall under the jurisdiction of research and educational organizations, their parent organizations bear institutional responsibility for enabling effective analysis, monitoring and enforcement of privacy practices across all phases of big data collection, compilation, analysis and application.

Our investigation on developing ethical guidelines for big data research determined that privacy needs to be treated relationally; that is, in context, with respect to the full range of potential ethical implications and with an eye to long term implications of recombination, reuse, reanalysis and repurposing. Making ethical decisions about data collection, documentation, and dissemination cannot thus be the product of a set formula. Researchers need to engage in a process of argumentative analysis of each set of choices for each context and each type of action. Ethical reasoning proceeds in the way of an argumentative analysis which formulates the terms of a possible decision as reasons favoring or questioning a specific data project in light of specific dimensions of ethical conduct, namely, beneficence, non-maleficence, autonomy, justice, and trust (see also Steinmann, Matei and Collmann in chapter 1). Again, no set formula can be offered as a magic bullet solution for determining the threshold over which an act can be considered ethical. However, it is important that reasons favoring a project should at least satisfy a majority of the criteria. In addition, failure to meet specific criteria should not create moral quandaries that are insurmountable at close scrutiny.

Upon completing this analysis, investigators should consider a final trade-off between sensitivity and tractability. In other words, what level of privacy protection does the specific data's sensitivity warrant? In this respect, one needs to keep in mind that protection bears material, intellectual and social costs. Protection involves limits and interdiction. Data that remain hidden from public view under multiple layers of safeguard are less likely to have socially beneficial effects. Protection of the most stringent kind might thus not be cost-effective, either materially or socially. If the sensitivity of the data is low, the process of protection becomes burdensome and rather intractable in the terms thus formulated. The investigator should, thus, make a decision that considers the overall impact of the privacy regime or strategy to be implemented. If the research offers multiple benefits, risks few negative, the sensitivity low and tractability of high stringency protection measures high, the investigator might decide to lower the level of protection. If sensitivity goes up and benefits down, the level of protection should increase, even if tractability becomes burdensome. The ethical reasoning process does not unfold deterministically. It requires considering each situation in its own right with respect to privacy context and ethical implications without automatically transferring assumptions between contexts.

4 Designing a Data Management Plan (DMP) for Big Data Projects

Several key insights significantly shape the DMP guidelines recommended here. First, it is important to recognize that while the scopes of privacy issues and information issues have some overlap, they are not the same. Hence, though information protections will be part of any DMP focus, they will not be sufficient to the task of formulating DMP guidelines for Common Rule institutions. These institutions will need specific privacy protections in addition to information management guidelines, including robust informed consent processes that go beyond IRB approval. This insight has also been communicated in the latest PCAST report (President's Council of Advisors on Science and Technology 2015: 50), as it states that the checkbox approach to information and privacy protections needs to be abandoned. Instead, Common Rule institutions should recognize that all those who interact with Big Data—whether in acquisition, storage, use and/or repurposing—share responsibility for the data, its sources and all uses to which it is put.

Second, in order to cover this broad spectrum of individuals who will be engaging Big Data from a diversity of interests and a variety of access points, DMP guidelines will need to foster the development of DMPs that establish, facilitate and reinforce Responsible Conduct in Research for any and all Big Data projects. Hence, the DMP will need to be an integral part of the entire Big Data research proposal, and will need to articulate how that integration will be accomplished, and by whom.

We emphasize that one size or type DMP will not fit all projects. After reviewing the terrain of Big Data research, it became clear that the broad scope of Big Data projects would require significant flexibility in any set of DMP guidelines in order to achieve the appropriate shape and size of DMP for a given project. Hence, the guidelines themselves must make this flexibility clear. This flexibility, however, does not lessen the responsibility of the investigators to craft a DMP that will achieve the level and scope of protections, and the depth of integration in the overall research project, that will be required by the guidelines. In order to articulate guidelines that will strike this balance between flexibility and rigor, we can take the approach of providing a list of questions for investigators to answer that will stimulate the creation of a robust DMP in such a way that reviewers will be able to ascertain with some ease the ways in which the DMP provides the protections and integration desired.

The opportunity to contemplate DMP guidelines that are essentially a response to the ELSI and concerns that arise from the growth of research, government, and business applications of Big Data is actually a part of the broader discussion of the purposes and use to which *any* data collected with the support of federal funds can and will be put. Individuals across disciplines and sectors who will be engaging Big Data from a diversity of interests and a variety of access points may benefit from guidance as to how data management planning can and should fit into their broader research plans. The experience of these researchers in their efforts to create good DMPs for Big Data research, with the heightened ethical issues it faces as described above in this chapter, may then benefit all researchers in their consideration of how best to manage data in any sized projects. Hence, the final objective of this effort is to foster the development of DMPs that establish, facilitate and reinforce Responsible Conduct in Research for any and all projects, whether or not the data they seek to curate, manage, collect or derive are considered "Big". In addition, especially for Big Data, but also in general, we recommend that the DMP needs to be an integral part of any research proposal. Considering the commonalities across research "with data" underscores that one size cannot possibly fit all when it comes to the DMP. Each Common Rule institution (or other funder) may have its own specific policies and those must also be integrated into the DMP; however, the DMP guidelines listed below have three key elements that every project can address explicitly. In keeping with our focus on the full and formal integration of data considerations (made explicit in the DMP) into the science, and scientific value, of the proposal, the answers to the prompts below should be derived directly from the research plan itself, i.e., we propose that data considerations should be formally integrated into the evaluation of the intellectual merit and broader impact of research funded by Common Rule agencies.

5 DMP Construction: Outline of a Decision Tree

Every DMP must contain information about the data (element 1); how it will be stored, housed, or managed during and beyond the funding period (element 2); and how access to the data will be managed, granted, rescinded, and otherwise controlled during and beyond the funding period (element 3). For the DMP, the data to which each element refers is the data on/from which investigators will be drawing inferences, and/or doing new analyses. We have developed a decision tree to help guide investigators through each element of the DMP. The decision tree invites

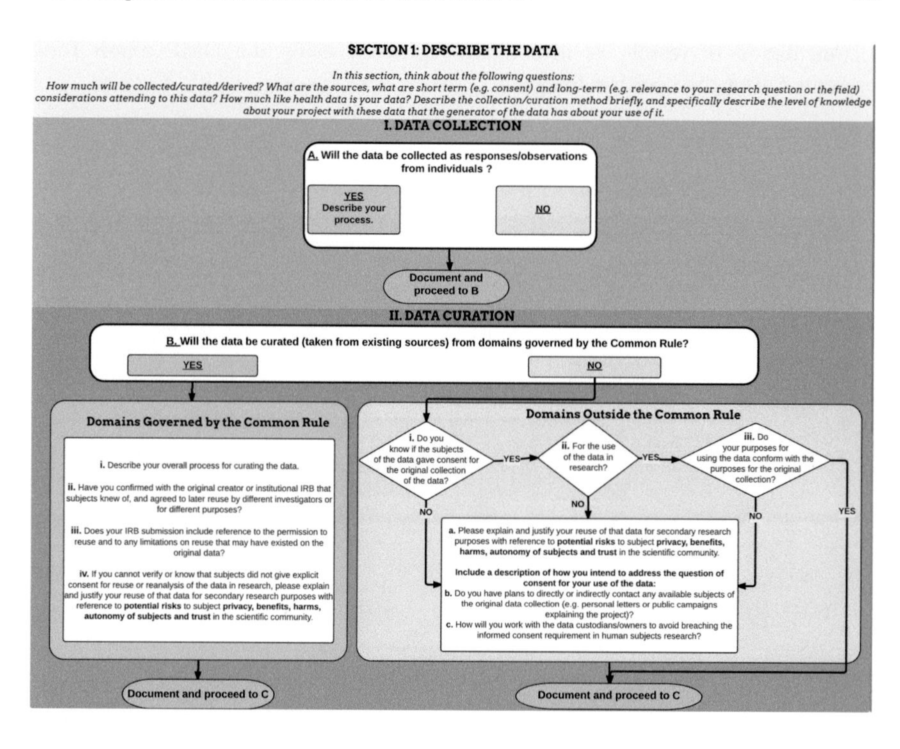
SECTION 1: DESCRIBE THE DATA
In this section, think about the following questions:
How much will be collected/curated/derived? What are the sources, what are short term (e.g. consent) and long-term (e.g. relevance to your research question or the field) considerations attending to this data? How much like health data is your data? Describe the collection/curation method briefly, and specifically describe the level of knowledge about your project with these data that the generator of the data has about your use of it.
I. DATA COLLECTION
A. Will the data be collected as responses/observations from individuals ?
YES Describe your process.
NO
Document and proceed to B
II. DATA CURATION
B. Will the data be curated (taken from existing sources) from domains governed by the Common Rule?
YES
NO
Domains Governed by the Common Rule
i. Describe your overall process for curating the data.
ii. Have you confirmed with the original creator or institutional IRB that subjects knew of, and agreed to later reuse by different investigators or for different purposes?
iii. Does your IRB submission include reference to the permission to reuse and to any limitations on reuse that may have existed on the original data?
iv. If you cannot verify or know that subjects did not give explicit consent for reuse or reanalysis of the data in research, please explain and justify your reuse of that data for secondary research purposes with reference to potential risks to subject privacy, benefits, harms, autonomy of subjects and trust in the scientific community.
Domains Outside the Common Rule
i. Do you know if the subjects of the data gave consent for the original collection of the data?
ii. For the use of the data in research?
iii. Do your purposes for using the data conform with the purposes for the original collection?
YES
NO
a. Please explain and justify your reuse of that data for secondary research purposes with reference to potential risks to subject privacy, benefits, harms, autonomy of subjects and trust in the scientific community.
Include a description of how you intend to address the question of consent for your use of the data:
b. Do you have plans to directly or indirectly contact any available subjects of the original data collection (e.g. personal letters or public campaigns explaining the project)?
c. How will you work with the data custodians/owners to avoid breaching the informed consent requirement in human subjects research?
Document and proceed to C
Document and proceed to C

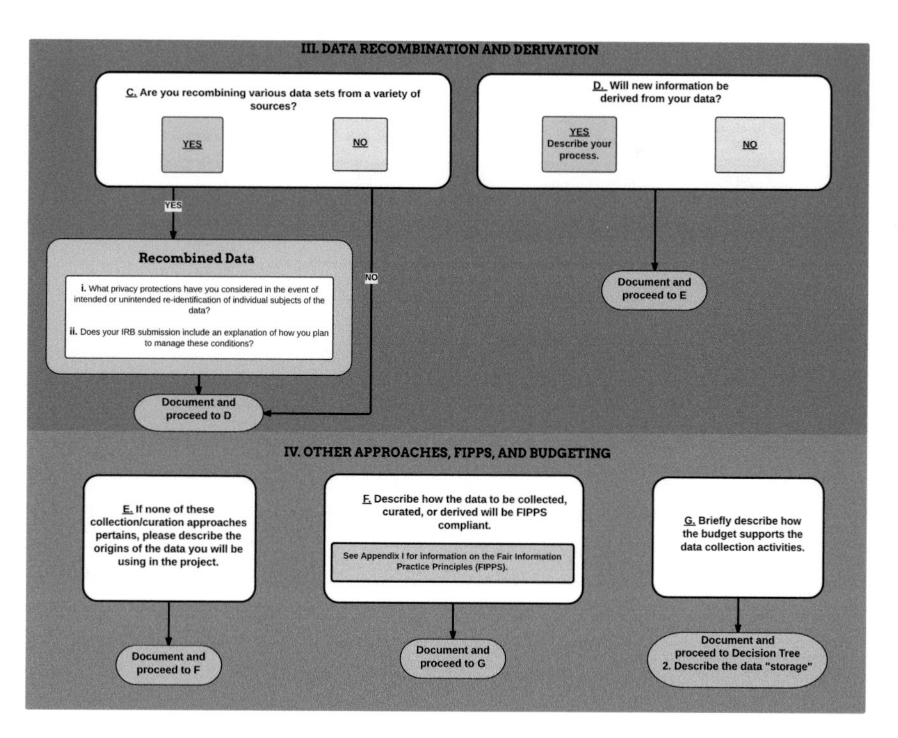
III. DATA RECOMBINATION AND DERIVATION
C. Are you recombining various data sets from a variety of sources?
YES
NO
D. Will new information be derived from your data?
YES Describe your process.
NO
Recombined Data
i. What privacy protections have you considered in the event of intended or unintended re-identification of individual subjects of the data?
ii. Does your IRB submission include an explanation of how you plan to manage these conditions?
Document and proceed to D
Document and proceed to E
IV. OTHER APPROACHES, FIPPS, AND BUDGETING
E. If none of these collection/curation approaches pertains, please describe the origins of the data you will be using in the project.
F. Describe how the data to be collected, curated, or derived will be FIPPS compliant.
See Appendix I for information on the Fair Information Practice Principles (FIPPS).
G. Briefly describe how the budget supports the data collection activities.
Document and proceed to F
Document and proceed to G
Document and proceed to Decision Tree 2. Describe the data "storage"

documentation of results as investigators proceed using the DMP report form (presented after the decision tree).

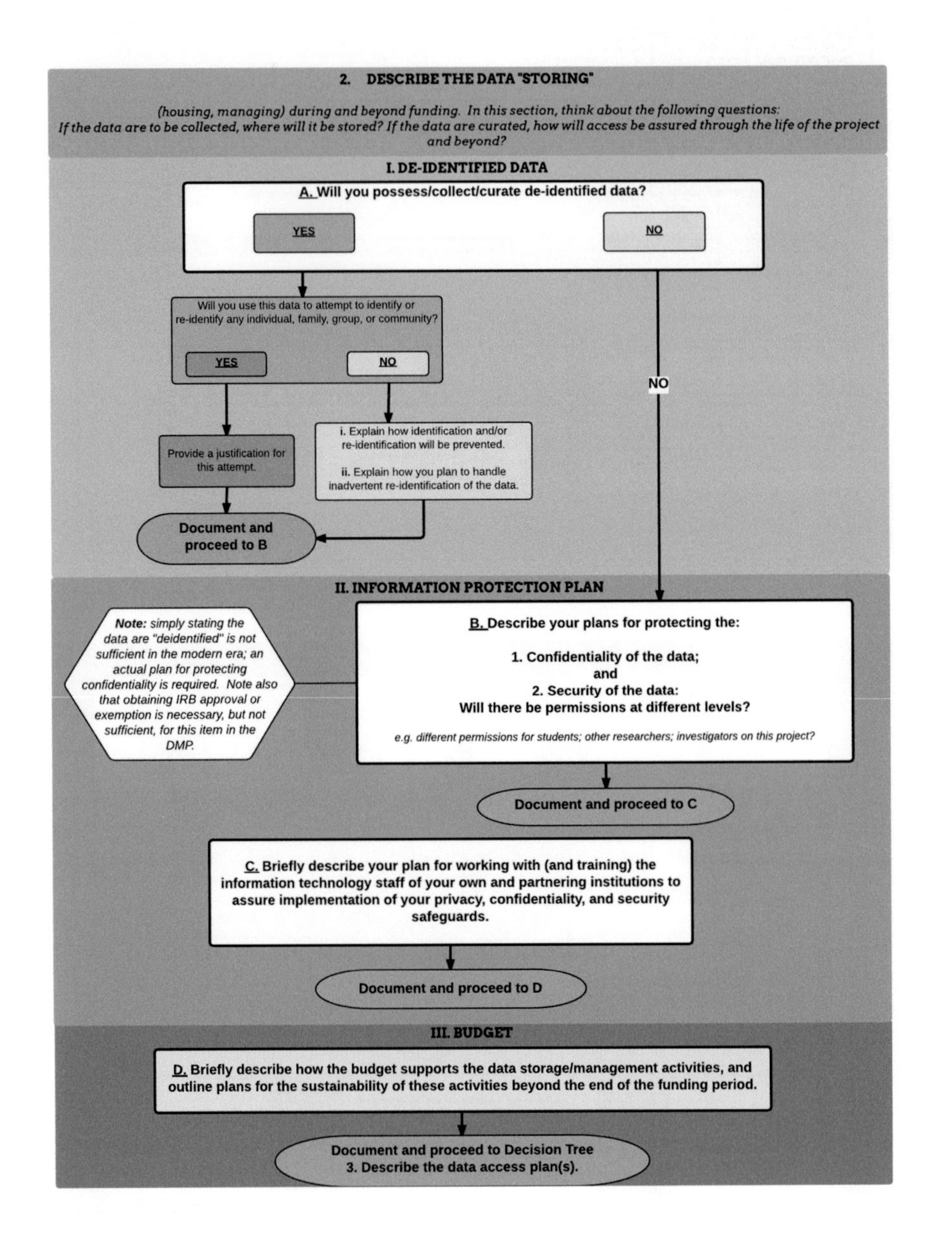

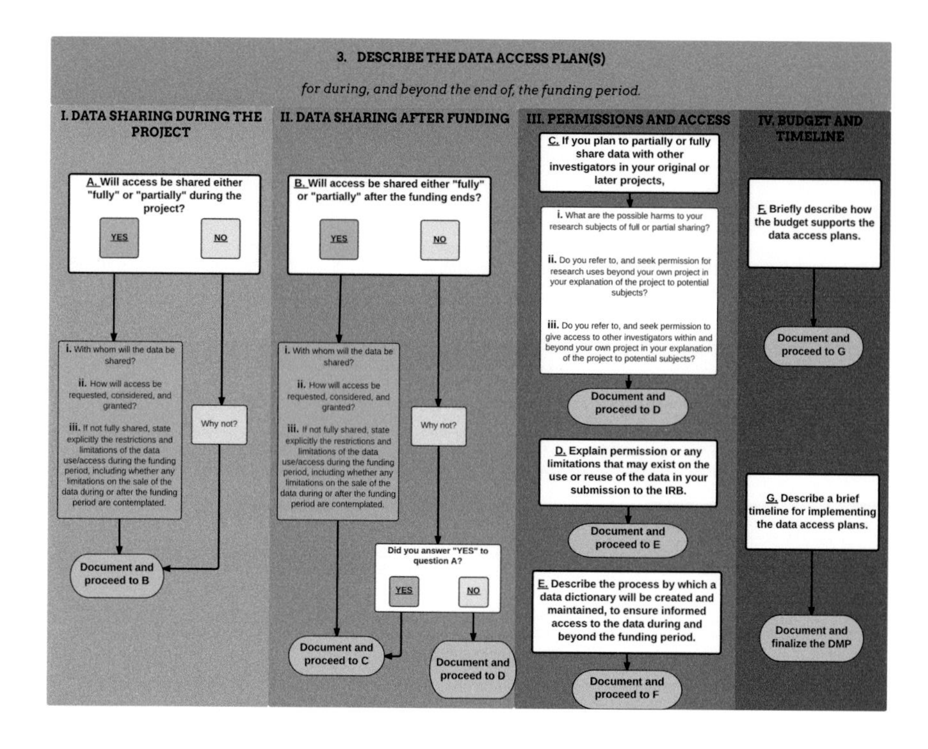
3. DESCRIBE THE DATA ACCESS PLAN(S)
for during, and beyond the end of, the funding period.
I. DATA SHARING DURING THE PROJECT
II. DATA SHARING AFTER FUNDING
III. PERMISSIONS AND ACCESS
IV. BUDGET AND TIMELINE
A. Will access be shared either "fully" or "partially" during the project?
YES
NO
i. With whom will the data be shared?
ii. How will access be requested, considered, and granted?
iii. If not fully shared, state explicitly the restrictions and limitations of the data use/access during the funding period, including whether any limitations on the sale of the data during or after the funding period are contemplated.
Why not?
Document and proceed to B
B. Will access be shared either "fully" or "partially" after the funding ends?
YES
NO
i. With whom will the data be shared?
ii. How will access be requested, considered, and granted?
iii. If not fully shared, state explicitly the restrictions and limitations of the data use/access during the funding period, including whether any limitations on the sale of the data during or after the funding period are contemplated.
Why not?
Did you answer "YES" to question A?
YES
NO
Document and proceed to C
Document and proceed to D
C. If you plan to partially or fully share data with other investigators in your original or later projects,
i. What are the possible harms to your research subjects of full or partial sharing?
ii. Do you refer to, and seek permission for research uses beyond your own project in your explanation of the project to potential subjects?
iii. Do you refer to, and seek permission to give access to other investigators within and beyond your own project in your explanation of the project to potential subjects?
Document and proceed to D
D. Explain permission or any limitations that may exist on the use or reuse of the data in your submission to the IRB.
Document and proceed to E
E. Describe the process by which a data dictionary will be created and maintained, to ensure informed access to the data during and beyond the funding period.
Document and proceed to F
F. Briefly describe how the budget supports the data access plans.
Document and proceed to G
G. Describe a brief timeline for implementing the data access plans.
Document and finalize the DMP

APPENDIX I: Fair Information Practice Principles

Five core principles of privacy addressed by FIPPS

Source: https://en.wikipedia.org/wiki/FTC_Fair_Information_Practice, United States Federal Trade Commision

Notice/Awareness

Consumers should be given adequate notice of an entity's information practices before any personal information is collected from them. This requires that companies explicitly notify some or all of the following:

- Identification of the entity collecting the data
- Identification of the uses to which the data will be put
- Identification of any potential recipients of the data
- The nature of the data collected and the means by which it is collected
- Whether the provision of the requested data is voluntary or required
- The steps taken by the data collector to ensure the confidentiality, integrity, and quality of the data

Choice/Consent

Choice and consent in an online information-gathering sense means giving consumers options to control how their data is used. Choice relates to secondary uses of information beyond the immediate needs of the information collector to complete the consumer's transaction. The two typical types of choice models are 'opt-in' or 'opt-out.' Each of these can be designed to allow a consumer to tailor the information gatherer's use of the information to fit their preferences by checking boxes to grant or deny permission for specific purposes rather than using a simple "all or nothing" method.

Opt-In

- Consumers affirmatively give permission for their information to be used for other purposes
- Information gatherer assumes that it cannot use the information for any other purpose without affirmation

Opt-Out

- Consumers must affirmatively decline permission for other uses
- Information gatherer assumes that it can use the consumer's information for other purposes with affirmation

Access/Participation

Access as defined in the FIPPS includes not only a consumer's ability to view the data collected, but also to verfiy and contest its accuracy. This access must be inexpensive and timely in order to be useful to the consumer.

Integrity/Security

Information collectors should ensure that the data they collect is accurate and secure. They can improve the integrity of data by cross-referencing it with only reputable databases and by providing access for the consumer to verify it. Information collectors can keep their data secure by protecting against both internal and external security threats. They can limit access within their company to only necessary employees to protect against internal threats, and they can use encryption and other computer-based security systems to stop outside threats.

Enforcement/Redress

In order to ensure that companies follow the Fair Information Practice Principles, there must be enforcement measures. The FTC identified three types of enforcement measures: self-regulation by the information collectors or an appointed regulatory body; private remedies that give civil causes of action for individuals whose information has been misused to sue violators; and government enforcement that can include civil and criminal penalties levied by the government.

Data Management Plan

Report Form

For use with the DMP Decision Tree

Section 1: Describe the Data

A. Will the data be collected as responses/observations from individuals?

   a. Yes
   b. No

B. Will the data be curated (taken from existing sources) from domains governed by the Common Rule?

   a. Yes
   b. No

C. Are you recombining various data from a variety of sources?

   a. Yes
   b. No

D. Will new information be derived from the data?

   a. Yes
   b. No

E. If none of these collection/curation approaches pertains, please describe the origins of the data you will use in the project.
F. Describe how the data to be collected, curated or derived will be FIPPS compliant.
G. Briefly describe how the budget supports the collection activities.

Section 2: Describe the data "storing"

A. Will you possess/collect/curate de-identified data?

   a. Yes
   b. No

B. Describe your plans for protecting the

   a. confidentiality of the data
   b. security of the data
   c. varying access permission levels

C. Briefly describe your plan for working with, and (when necessary) training the information technology staff of your own and partnering institutions to assure implementation of your privacy, confidentiality, and security safeguards.
D. Briefly describe how the budget supports the data storage/management activities, including activities beyond the funding period.

Section 3: Describe the data access plans

A. Will access to the data be shared either fully or partially during the project?

   a. Yes
   b. No

B. Will access to the data be shared either fully or partially after the funding ends?

   c. Yes
   d. No

C. What are the possible harms to your research subjects of full or partial sharing?
D. Explain permission or any limitations that may exist on the use or reuse of the data in your submission to the IRB?
E. Describe the process by which a data dictionary will be created and maintained
F. Briefly describe how the budget supports the data access plans
G. Briefly describe the timeline for implementing data access plans.

6 Application of the DMP to Specific Cases

Presented below are five cases that demonstrate how to use the DMP decision tree and the accompanying DMP report form for human subjects research in which big data is recombined, repurposed, reused, or reanalyzed. Each section includes a brief summary of the case, a walk-through of the decision tree and report form, and commentary.

6.1 Case 1: Recombining Existing Data for New Research

Data Management Plan
Report Form
For use with the DMP Decision Tree
Investigators at Georgetown University are recombining multiple sources of big data about forced migration to better forecast, respond to, and help alleviate the consequences of humanitarian crises (Berkowitz et al. 2014; Collmann et al. 2014a, b; Hirschhorn 2014). Because detailed local data is difficult to obtain in a timely manner, this project explores the effectiveness of using open-source, online data to help identify indirect, leading indicators of displacement/forced migration. Indicators relevant to this project include: economic, political, social, demographic and environmental changes affecting movements; intervening factors such as government refugee policies; and community and household characteristics. Parsing irrelevant information from the true indicators, calibrating results, understanding how these indicators change through time, and identifying and removing

potential bias, requires large-scale data analysis and potentially, new computational methods for developing meaningful descriptive and predictive models. To date, the big data Georgetown uses for this study include open-source media articles and Twitter data. Georgetown has access to EOS, a vast unstructured archive of over 700 million publicly available open-source media articles that has been actively compiled since 2006. New articles are added at the rate of approximately 300,000 per day by automated scraping of over 11,000 Internet sources in 46 languages across the globe. Georgetown also collects data from Twitter—hundreds of thousands of tweets per day for the last 6 months. When relevant, we also draw data from the scholarly literature of history, anthropology, economics and other social sciences as well as the gray literature of governmental and non-governmental organizations. Long-term plans include adding data from the archives of collaborating international and non-governmental organizations.

This case highlights the importance of protecting the results of big data analyses with the explicit intent of better describing and aiding specific individuals or communities; that is, potentially re-identifying people and, thus, creating privacy breaches for the purpose of humanitarian aid. In addition, investigators will conduct interviews with select samples of displaced persons.

Section 1: Describe the Data

A. Will the data be collected as responses/observations from individuals?

   a. Yes: We will conduct interviews with select samples of displaced persons. The project calls for conducting interviews with select samples of displaced persons to obtain accounts of their individual displacement motivations, experiences, and developmental histories and, thus, enable comparison with data gathered from open sources.
   b. No

B. Will the data be curated (taken from existing sources) from domains governed by the Common Rule?

   a. Yes
   b. No: the data has already and will be collected primarily from open sources such as media articles and accessible social media (commercial contexts). Other potential sources include official reports of governmental organizations, international governmental organizations, and non-governmental organizations (governmental, non-governmental contexts). Consent practices vary across this range of data sources. Journalists may or may not announce their reporting intentions to people about whom they write, depending on the method by which they obtain their information. Journalists also may or may not identify their human sources also depending on the circumstances. The reports of governmental, international governmental and non-governmental organizations typically appear in the form of structured enumeration of attributes of populations rather than of individual persons. We have used and intend to continue using open source data for better

understanding and potentially forecasting the dynamics of forced migration, including detailed descriptions of who, when and where identified individuals and group become displaced.

i. Benefits: the research project derives its purpose from current failures in adequately responding to the needs of potentially or actually displaced persons during crises. In spite of years of practical experience, governments and relief agencies routinely fail to make accurate forecasts about the numbers, destinations or timing of refugee movements even when given knowledge of imminent crises such as an expected invasion. It is hoped that an improved forecasting tool will enhance the overall effectiveness and adequacy of relief efforts and, thus, a reduction in the suffering, morbidity and mortality of refugees.
ii. Harm: not all stakeholders share the same interest in the fate of refugees during a crisis, especially but not only in conflict situations. As recent developments in the Middle East and Europe indicate, large flows of refugees potentially elicit hostile responses even from non-combatant governments because of their potential impact on the resources and political dynamics of host nations. In war zones, knowing the identities and intentions of refugees and their leaders may lead to direct harm such as arrest, military targeting, forced encapsulation, and other coercive measures.
iii. Autonomy of subjects: For us, the ethical analysis of the autonomy of research subjects under conditions of forced migration turns on the mission of the project and the method of data acquisition. Persons undergoing forced migration often suffer diminished autonomy even as they exercise their best judgment in attempting to mitigate the threatening the conditions they face. When effectively conducted, the mission of humanitarian relief aids in restoring some measure of refugee autonomy as it protects them from violence, forestalls starvation, and mitigates the threat of infectious disease. Developing tools to enhance effectiveness of humanitarian relief, thus, potentially also enhances refugee overall autonomy. As described above, when we interview refugees, we obtain informed consent. Using open source materials, however, poses different questions. The subjects of open media reporting have little control over the behavior of journalists beyond refusing to grant interviews and the laws governing false or libelous reporting. The social value of the press makes open media possible but distinguishes it as a domain from Common Rule research. The question of the consent status of social media remains open with some arguing that participation in conversations on social media such as Twitter constitutes public acts and, thus, entail implicit consent for any subsequent use. Others disagree (Zimmer and Proferes 2014). We have adopted the perspective that the overall mission of humanitarian relief and its potential for enhancing the autonomy of forced migrants in extreme circumstances warrants using both open source media and available social media as

sources of information about unfolding crises and their impact on populations. As described below, however, we create, maintain and protect the open source materials we collect for EOS and Twitter and, for that reason, take special precautions to protect the identity of persons identified in the research even as we acquire data about refugees over which they have little or no control.

C. Are you recombining various data from a variety of sources?

   a. Yes: At this point in the project, we use data from two primary open sources, the Georgetown-compiled EOS database and a selected sample of Twitter messages from the region of interest (Middle East). As the project develops, we intend to add other structured databases from collaborating organizations. The EOS database has a long history of protecting, and, in some cases, not reporting individual identities of persons appearing in open media (see Collmann and Robinson 2010). Project Argus had a public health biosurveillance mission that required only aggregate analyses of threats to populations not to individuals. Thus, it adopted and enforced a rule to never report individual names of persons inadvertently identified during analysis of open source materials.
   As described above, however, this project seeks and intends to use detailed identifiable information about individual refugees whenever possible and relevant. This may occur as a result of analyzing a variety of reports from open media, or social media. It is most likely to occur, however, when we analytically link the open source data with data from a global database derived from Landscan, a map of the global population distribution (http://web.ornl.gov/sci/landscan/). Landscan represents the world's entire population in the form of statistical human entities constructed annually from census reports and distributed as they are found on the earth at approximately 1 km resolution (30″ × 30″). We have already augmented Landscan with attributes about the statistical human entities from other sources such as the CIA Factbook with data about gender, religion, and other relatively static social attributes. When adding dynamic, relatively current data from open media reports and Twitter on geo-located events associated with civil conflict, natural disasters, disease outbreaks and other conditions promoting forced migration, the possibility arises of naming and locating persons who would otherwise remained unidentified and hidden.
   At this point in the project, we have just begun to explore the privacy and human subject implications of these possibilities. Important questions remain unanswered and lack data yet to begin finding answers, such as:

   - How often do previously unidentified persons become identified through the recombining of the various databases in the forecasting forced migration project?
   - What harm could come to such persons as a result of identification?

- Does recombining some databases rather than others elevate the risk of identification?
- How does the risk of harm from identification vary across the spectrum of situations being encountered through the course of displacement from initial flight to resettlement?

Answers to these questions might help address other pertinent questions, such as:

- Does the effectiveness of the forecasting tool vary according to the accuracy and detail in identifying specific individuals?
- How should we structure access to the data and to the results of analysis?

These unanswered questions encourage caution at this stage of the project. The mere fact that open source and social media data circulate among the public does not necessarily mean that the results of our analysis of the data pose minimal risk to the people or groups under study. Responsible conduct of this research entails close collaboration with the Georgetown IRB and university computer services during the planning, data acquisition, data analysis and reporting phases of the project. The project will include gathering data about the questions posed above and careful monitoring of emergent identification incidents.

b. No.

D. Will new information be derived from the data?

a. Yes: Over the course of the project, developing the forecasting tool will create and employ derived data (see 1B and 1C above)
b. No

E. If none of these collection/curation approaches pertains, please describe the origins of the data you will use in the project. Not applicable
F. Describe how the data to be collected, curated or derived will be FIPPS compliant.
Data gathered during interviews in this project fall subject to IRB review and will require authorization for any future reuse. Open source data fall outside the scope of FIPPS having never been collected for any specific purpose in the first place.
G. Briefly describe how the budget supports the collection activities.

a. The budget includes specific costs for salary, travel, interview fees, and data analysis for all interviews of refugees and humanitarian relief workers.
b. The budget includes support for the Georgetown University computer services staff that supports EOS and the social media collection activities.

Section 2: Describe the data "storing"

A. Will you possess/collect/curate de-identified data?

   a. Yes
   b. No: at this point in the project, we do not plan to collect de-identified data about individuals.

B. Describe your plans for protecting the

   a. confidentiality and security of the data: our team and our partners in this project will treat the results of project analysis as sensitive and, thus, subject to the range of security controls identified by Georgetown University information security policies as appropriate for sensitive data. The university information services staff responsible for building and maintaining EOS, the Twitter database, the enhanced Landscan data base, and any other computerized information resources that may become part of this project serve on the staff of this project and participate in all project staff meetings. The director of the university information services team supporting this project also serves on the project steering committee. The lead investigators from all collaborating institutions also serve on the project steering committee and, thereby, ensure joint discussion and common understanding of confidentiality and security issues as they arise in the project.
   b. varying access permission levels: Georgetown University grants access to EOS, the Twitter database and enhanced Landscan only to authorized university investigators or their research partners.

C. Briefly describe your plan for working with, and (when necessary) training the information technology staff of your own and partnering institutions to assure implementation of your privacy, confidentiality, and security safeguards.

   a. EOS: authorization to use EOS includes receiving training in gaining access and appropriately using the results of searches. Users learn and agree to respect the fair information practice of using EOS only for designated projects and, thus, not for reuse, sale or other forms of transfer to unauthorized projects or users. The Georgetown university information services staff helps enforce these policies through following policies about enrolling and auditing users.
   b. Twitter:
   c. Enhanced Landscan:

D. Briefly describe how the budget supports the data storage/management activities, including activities beyond the funding period.

   a. Georgetown University provides ongoing support for EOS, the Twitter database and enhanced Landscan as part of its support for academic computing.

b. The budget for this project supports the addition of new resources such as scraping new open media sources, Twitter and other online materials as required.

Section 3: Describe the data access plans

A. Will access to the data be shared either fully or partially during the project?

   a. Yes: we currently collaborate with investigators from several organizations in this project including Georgetown University, York University, Sussex University, Kultur University, and Lawrence Livermore National Laboratory. Depending on their role in the project, investigators and staff have access to different information resources.
   b. No

B. Will access to the data be shared either fully or partially after the funding ends?

   a. Yes: We give access to these information resources to authorized research collaborators only.
   b. No

C. What are the possible harms to your research subjects of full or partial sharing? In order to respect the principles of fair information use, we have policies that restrict sharing copyrighted open source media with our collaborators in research projects.
D. Explain permission or any limitations that may exist on the use or reuse of the data in your submission to the IRB.
   The IRB submission includes a description of EOS and of the limitations on its use.
E. Describe the process by which a data dictionary will be created and maintained
F. Briefly describe how the budget supports the data access plans.
   The budget includes funds for support of university information services to maintain the open source databases and for research assistants to help instruct investigators in their use.
G. Briefly describe the timeline for implementing data access plans.
   The project provides immediate access after training to any authorized user, including investigators, graduate students and other collaborators.

Commentary on Case 1: Recombining Existing Data for New Research

This type of project that conducts analyses over a range of data types and data sources with varying ethical provenances may constitute a significant proportion of big data projects in the social sciences and humanities. Debate centers particularly on the question of how best to handle unstructured, open source data such as media and Twitter. The investigators at Georgetown have adopted the perspective that we know too little about the potential privacy and ethical questions of open source data when addressing populations of people such as forced migrants to treat the data and, especially, the results of the analysis as purely public. This is true even though the data originates from the public sphere in the broad sense of the term. Thus, the research team plans to work closely with the IRB and has already established close collaboration with university information services to ensure that both the human subjects and the data receive appropriate protection. Because the investigators see the possibility of harm coming to persons through the results of data analysis, they refer to this dual concern with protection of both the human subjects and the data as data stewardship.

Our reading of the NPRM Section 2. Explicit Exclusion of Activities from the Common Rule leads us to believe, however, that the open source section of Georgetown's forced migration project, would be excluded from consideration as a human subjects protection issue under the proposed revisions. In paragraph 2.9. the NPRM defines criteria for exclusion that appear to apply to EOS, Georgetown's open source media database.

> Collection or Study of Information that has been or will be Collected; applies to research involving the use of existing data, documents, records, and pathological or diagnostic specimens, but only if the sources are publicly available or if the information is recorded by investigators in such a manner that subjects cannot be identified, directly or through identifiers linked to them. (Federal Register 2015: 53952)

In the terms of this paragraph, EOS is a collection of existing data and documents that are publicly available and not recorded in a manner that identifies human subjects. The paragraph offers the following logic for excluding such data from human subjects protection.

> The underlying logic behind the exclusion in proposed §__.101(b)(2)(ii) is that such research involves no direct interaction or intervention with human subjects, and any research use of the information does not impose any additional personal or informational risk to the subjects, because (1) the information is already available to the public, and so any risk it may include exists already, or (2) the information recorded by the investigator cannot be identified, and no connection to or involvement of the subjects is contemplated. Any requirements of the Common Rule would not provide additional protections to subjects, and could add substantial administrative burden on IRBs, institutions, and investigators. Creating this excluded category avoids that problem. (Federal Register 2015: 53945)

We believe that this logic fails because it addresses only the source of the data and not the potential harm of the results of data analysis. At a minimum, we suggest that Common Rule institutions lack sufficient experience with complex, highly synthesized big data resources to so quickly dismiss the risk their use may potentially pose to subjects, particularly subjects classically identified as vulnerable, or, more generally, subjects who become identified as subjects in the course of the analysis rather than at the start of the project. In such cases, consent of any kind becomes difficult if not impossible and the risk of harm is currently inestimable. We propose that such projects require full IRB review, ongoing monitoring and tight collaboration with university information services.

6.2 Case 2: Repurposing Administrative Data for Research

Data Management Plan

Report Form

For use with the DMP Decision Tree

Social science investigators are finding great value in linking administrative records from multiple administrative data systems/entities into longitudinal datasets at different levels of analysis (individual, family, program, school, etc.). State and local agencies collect the data for multiple purposes, such as program accountability, client tracking, and service effectiveness. "Data" in this case refers to administrative records, established when a person or family applies for social, health or educational services (such as enrolling a child in school). Most states also routinely collect data on newborn babies, their parents (especially mothers), including prenatal care; any birth defects or signs of vulnerable health (e.g. hearing loss registry); and other important social indicators. Public health departments collect a range of data and routinely work with hospitals, clinics and other providers in tracking persons to ensure adequate care and provision of services as well as effective disease monitoring. The following DMP is for a generic case of repurposing administrative data for research, intended to account for state-by-state and departmental variation in how records are collected, stored and protected. Some responses will have to be tailored to your specific research project, as indicated by text in italics.

Section 1: Describe the Data

A. Will the data be collected as responses/observations from individuals?

   a. Yes
   b. No: a state/local agency or a public health department has previously collected the data

B. Will the data be curated (taken from existing sources) from domains governed by the Common Rule?

   a. Yes
   b. No: State and local agencies and public health departments fall outside the Common Rule

i. [*For data stored in state/local agencies and public health departments, it may be unclear as to whether subjects consented to the original collection of the data or to use of the data in research. This varies by state, and possibly by department.*]

iv. [*Explanation and justification of reuse of the data will depend on the particular research project and its purpose.*]

Benefits: *Again, this will depend on the goals of the research project.*

Privacy: *The particular privacy risks posed to individuals involved will depend on the research project. However, it is important to consider whether you are working with personally identifiable data or de-identified data, both of which have particular risks associated with them. The consent status of the data (i.e. whether subjects consented to the original collection of the data or to use of the data in research), and the implications of moving data from one context (government) to another (Common Rule institution) become relevant and important to consider.*

Harms: *These will be specific to the research project, but can be potentially physical, mental, emotional, social, economic, or legal harms implicated in your research for participating subjects.*

Autonomy of subjects: *Response depends upon the research project. The consent status of the data becomes relevant here, as well.*

Trust in the scientific community: *Potential risk to trust in the scientific community depends upon the specifics of the research project. Consider: trust can be put at stake, particularly if there is accidental or unauthorized disclosure of information about an individual or a particular population, and the subjects had no knowledge that the data was (a) originally collected and that (b) the data was being used in research that had a purpose other than for what it was originally collected.*

iv. a and b. If you do not know if the subjects gave consent for the original collection of the data, it is recommended that you consult with the original data collectors to determine the status of consent for collection. From there, you may determine if it is necessary to contact subjects of the original data collection.

C. Are you recombining various data from a variety of sources?
[*Response will depend upon the specifics of the research project.*]

  a. Yes

    i. *If data will be recombined, we encourage consultation with computer security staff, institutional IRBs, and data owners/custodians to ensure that the transfer of data, the use/study of data within your own institution, and the storage of such data will involve appropriate safeguards for privacy and information protection for subjects.*

  b. No

D. Will new information be derived from the data?
[*Response will depend upon the specifics of the research project.*]

  a. Yes
  b. No

E. If none of these collection/curation approaches pertains, please describe the origins of the data you will use in the project.

F. Describe how the data to be collected, curated or derived will be FIPPS compliant (See Appendix I: FIPPS).
Notice/Awareness: [*Response will depend upon the consent status of the data and the specifics of the research project.*]

Choice/Consent: Subjects will be able to opt-out of this study, if they do not wish their original data to be used for research.
Access/Participation: [*Response will depend upon state/departmental regulations and specifics of the research project*]
Integrity/Security: [*Response will depend upon state/departmental regulations and the specifics of the research project, but consultation with information technology staff and computer security staff is recommended*]
Enforcement/Redress: [*Response will depend upon state/departmental measures and specifics of the research project, but it is recommended that research staff be trained appropriately using institutional resources and regulatory bodies (e.g. Office of Ethics and Compliance, IRB)*]

G. Briefly describe how the budget supports the collection activities.
[*Consider items such as: consultation (with IRBs, computer security staff, information technology staff, and subjects), provision of notice and informed consent forms for subjects, and/or in soliciting informed consent from subjects,* etc.]

Section 2: Describe the data "storing"

B. Will you possess/collect/curate de-identified data?
[*Response will depend upon the type of data and specific goals of the research project.*]

  a. Yes
  b. No

C. Describe your plans for protecting the

  a. confidentiality of the data: [*depends upon the specifics of the research project*]
  b. security of the data: [*depends upon the specifics of the research project, but consultation with computer security staff is recommended to ensure that the data is appropriately protected (both digitally and physically) from unauthorized access, use, and disclosure*]
  c. varying access permission levels: [*depends upon the specifics of the research project and state, local, and/or departmental regulations*]

D. Briefly describe your plan for working with, and (when necessary) training the information technology staff of your own and partnering institutions to assure implementation of your privacy, confidentiality, and security safeguards.
[*This will depend upon the specifics of the research project. Consider: ensuring that investigators and staff are informed about any sensitive information, analytic data, and the implications of unauthorized use or disclosure of personally identifiable information and the risks that these pose to human research subjects; consultation with institutional resources and computer security staff.*]

E. Briefly describe how the budget supports the data storage/management activities, including activities beyond the funding period.
[*Response will depend upon specifics of the research project.*]

Section 3: Describe the data access plans

A. Will access to the data be shared either fully or partially during the project? [*Response will depend upon the goals of the research project.*]

   a. Yes
   b. No

B. Will access to the data be shared either fully or partially after the funding ends? [*Response will depend upon goals of the research project.*]

   a. Yes
   b. No

C. What are the possible harms to your research subjects of full or partial sharing? [*Response will depend upon goals of the research project. Be sure to consider potential physical, emotional, mental, social, economic, and legal harms posed by the research project to participating subjects and any groups to which they might belong*]
D. Explain permission or any limitations that may exist on the use or reuse of the data in your submission to the IRB.
[*Response will depend upon specifics of the research project.*]
E. Describe the process by which a data dictionary will be created and maintained [*Response will depend upon specifics of the research project. If necessary, you may consult with institutional resources or information technology staff.*]
F. Briefly describe how the budget supports the data access plans.
[*Response will depend upon specifics of the research project.*]
G. Briefly describe the timeline for implementing data access plans.
[*Response will depend upon specifics of the research project.*]

Commentary on Case 2: Repurposing Administrative Data for Research Case
Case 2 brings attention to the unique privacy and information protection risks raised by repurposing data: moving data that was originally collected for a specific purpose in one domain (government) to another domain (Common Rule institution) in which it will be analyzed for unrelated purposes (research). Ethical provenance becomes relevant in this type of research, because the conditions under which the data were originally collected have implications for new and future uses of such data. Specifically, sometimes records are used differently from their original purpose, such as cases in which records were originally collected without consent for use in research and potentially are being repurposed without the knowledge of individuals that the records concerned. Such uncertainty in the consent status of the data poses risks to privacy, subject autonomy, and trust. These risks become evident in Sect. 1.B.i.iv. of the DMP Decision Tree, where the investigator is asked to justify use of the data for secondary research purposes with regard to the fundamental principles. Upon completion of the DMP for Case 2, it becomes clear that the investigator needs to take the appropriate steps to determine the consent status

of the data, including how consent was obtained and for what specific purposes. Additionally, the DMP Decision Tree encourages the investigator to consider potential harms to individuals involved in their study, such as breaches of privacy and information, and thus, to consider provision of commensurate safeguards to protect both. It further encourages investigators to seek guidance from IRBs and computer security staff at their respective institutions.

The recently released Common Rule NPRM discusses autonomy interests implicated in the secondary use of biospecimens, and suggests that such interests also apply to research involving identifiable private information:

> Regardless of the scale on which harms may have occurred in the past, continuing to allow secondary research with biospecimens collected without consent for research places the publicly funded research enterprise in an increasingly untenable position because it is not consistent with the majority of the public's wishes, which reflect legitimate autonomy interests. (Federal Register 2015: 53944).

The NPRM also proposes exempt reviews for secondary use of identifiable private information:

1. iii. Secondary Research Use of Identifiable Private Information (NPRM at §__.104(e)(2))
    a. Prior notice has been given to the individuals to whom the identifiable private information pertains that such information may be used in research;
    b. The privacy safeguards of §__.105 are required; and
    c. The identifiable private information is used only for purposes of the specific research for which the investigator or recipient entity requested access to the information. (Federal Register 2015: 53963)
2. Exemption for the Storage or Maintenance of Biospecimens or Identifiable Private Information for Secondary Research Use (NPRM at §__.104(f)(1)) if the following criteria are met:

    Written consent for the storage, maintenance, and secondary research use of the information or biospecimens is obtained using the broad consent template that the Secretary of HHS will develop. Oral consent, if obtained during the original data collection and in accordance with the elements of broad consent outlined in §__.116(c) and (d)(3), would be satisfactory for the research use of identifiable private information initially acquired in accordance with activities excluded under §__.101(b)(2)(i) or exempt in accordance with §__.104(d)(3) or (4), or §__.104(e)(1); and The reviewing IRB conducts a limited IRB review of the process through which broad consent will be sought, and, in some cases, of the adequacy of the privacy safeguards described in §__.105.
    Note: This exempt category is for secondary research use of biospecimens and identifiable private information and applies to biospecimens and identifiable private information that were initially collected for purposes other than the proposed research activity. The term 'other than the proposed activity' here

means that the information or biospecimens were or will be collected for a different research study or for non-research purposes. (Federal Register 2015: 53966)

The NPRM, in an attempt to balance the "legitimate autonomy interests" against the challenges of obtaining consent on a large scale, includes the proposal of broad consent. The tenets of and the actual forms for this broad consent mechanism have yet to be published and critiqued, and so much of the NPRM approach will depend on successfully developing these forms and processes.

Repurposing the administrative data collected by state/local agencies or departments for research holds many potential benefits, particularly for social science, public health, and epidemiological research. However, repurposing big data, especially when such data is traced across time and spheres of life and used to create personal profiles, can pose significant risks to privacy for both groups and individuals. These risks necessitate careful consideration of the consent status of the original data and potential harms posed to research subjects. As the NPRM's broad consent mechanism remains to be seen, we cannot be certain that it will encourage such careful consideration. Our DMP Decision Tree and process, as demonstrated above, do indeed help the investigator to consider the unique privacy and information protection risks posed by repurposing administrative data for research, namely, the consent status of the original data and the potential harms implicated in their research. Particularly as technology advances, data sets grow in size, and our ability to aggregate, analyze, and move large data sets between contexts improves, regard for these risks to subject autonomy, privacy, and trust becomes increasingly relevant and important.

6.3 *Case 3: Repurposing Newborn Screening Data*

Data Management Plan

Report Form

For use with the DMP Decision Tree

State mandated programs provide screening and data collection for 4 million newborns in the U.S. each year. After newborn screening is completed, the residual dried blood spots (RDBS) and data can be stored for quality assurance and research purposes depending on state practice and statutes. Currently, fourteen states store RDBS for research purposes. Storage of RDBS and data can range from a few months to the entire life of the program depending on the state. For example, California has stored the RDBS and data for newborns born in California for the past 52 years. Programs in Minnesota and Texas lost law suits alleging use of their RDBS for purposes not included in the original parental consent, including but not limited to research. Indiana is currently involved in an active lawsuit (see "The Ethics of Large-Scale Genomic Research" by Berkman et al. for a detailed description of the program and its past legal problems). More recently, in

December, 2014, the Newborn Screening Saves Lives Reauthorization Act was signed into law. This law changed the status of newborn blood spots with regards to research by eliminating the option of IRBs to waive consent requirements for research with the blood spots as they were now to be considered research with human subjects. The following DMP is for a generic case of repurposing Newborn Screening (NBS) data for research. Some responses will have to be tailored to your specific research project, as indicated by text in italics.

Section 1: Describe the Data

A. Will the data be collected as responses/observations from individuals?

   a. Yes
   b. No: the study will involve the use of blood spots that were previously collected by state governments as part of their Newborn Screening programs.

B. Will the data be curated (taken from existing sources) from domains governed by the Common Rule?

   a. Yes
   b. No: the source of the data will be from the blood spot repositories managed by each participating state government.

C. Are you recombining various data from a variety of sources?

   a. Yes
   b. No: all the data will come from the state biorepositories.

D. Will new information be derived from the data?

   a. Yes: though the exact research projects are not yet delineated, the data may be used to do additional research on diseases in individuals already identified by newborn screening program, or to look for other indications of genetic disease or difference.
   b. No

E. If none of these collection/curation approaches pertains, please describe the origins of the data you will use in the project. N/A
F. Describe how the data to be collected, curated or derived will be FIPPS compliant.
   Notice: In the regulations for a given state's newborn screening program there are guidelines regarding the rights parents have for opting out of any future research program after the spots are no longer needed for screening purposes. Information about this process is available to parents through public notice by the individual state governments. Also included in this information are details regarding the length of time blood spots may be retained in a secure biorepository, and how the blood spots may be reanalyzed in the future if circumstances warrant.

Consent: The consent process for this program is usually an opt out process, with the parents able to choose not to have their newborn's blood spots used for research projects outside of the screening program.

Access: States may vary in the access parents (and newborns later in life) will have to the blood spots and any derived data.

Security: each State takes measures (those these vary among the states) to protect the security of the newborn blood spots and medical data, and who has what access to these materials for approved research projects. Researchers using newborn screening data need to negotiate security procedures and requirements with the States whose data they are using, and follow any Common Rule requirements if applicable.

Enforcement: each State is responsible for enforcement of its regulations and guidelines—which may vary among the states. Researchers may also need to follow Common Rule requirements if applicable.

G. Briefly describe how the budget supports the collection activities.

N/A—the blood spots have already been collected as part of the state Newborn Screening program.

Section 2: Describe the data "storing"

A. Will you possess/collect/curate de-identified data? *Each research project will have to answer this question in light of the focus of the research.* [*E.g. QA research may be de-identified, while disease research may not.*]

  a. Yes
  b. No

B. Describe your plans for protecting the

  a. confidentiality of the data
  b. security of the data
  c. varying access permission levels

  [*Specific answers to these questions will depend on a state's guidelines and the structure and focus of each research project.*]

C. Briefly describe your plan for working with, and (when necessary) training the information technology staff of your own and partnering institutions to assure implementation of your privacy, confidentiality, and security safeguards.

  [*Specific answers to these questions will depend on the specific features of each research project.*]

D. Briefly describe how the budget supports the data storage/management activities, including activities beyond the funding period.

  [*Specific answers to these questions will depend on the specific features of each research project.*]

Section 3: Describe the data access plans

A. Will access to the data be shared either fully or partially during the project? [*Answer will depend on specific goals of each research project.*]

   a. Yes
   b. No

B. Will access to the data be shared either fully or partially after the funding ends? [*Answer will depend on specific goals of each research project.*]

   a. Yes
   b. No

C. What are the possible harms to your research subjects of full or partial sharing? [*The answer to this question should include potential physical, mental, emotional, social, economic or legal harms the research subjects may encounter.*]
D. Explain permission or any limitations that may exist on the use or reuse of the data in your submission to the IRB. [*Document this part of your IRB submission.*]
E. Describe the process by which a data dictionary will be created and maintained [*Answer will be specific to each research project.*]
F. Briefly describe how the budget supports the data access plans [*Answer will be specific to each research project.*]
G. Briefly describe the timeline for implementing data access plans. [*Answer will be specific to each research project.*]

Commentary on Case 3: Repurposing of Biospecimens (Blood Spots) for Use in Research Projects Outside of the Scope of the Newborn Screening Program

The value and success of the Newborn Screening Program in the U.S. is widely acknowledged and applauded. This tremendous social benefit creates an equally tremendous obligation on the part of the individual States and the Federal government to insure that this program continues to provide adequately funded, evidence based, newborn disease screening relevant to each State's population. Hence, any use of the stored blood spots, and/or derived health status or genetic testing information, which is outside of the scope of the screening program must be evaluated in terms of any risk or detriment to the program that use of the biospecimens and information might incur. The recent removal of the informed consent waiver option for IRBs for research involving newborn blood spots by the passage of the Reauthorization Act is an indication of the Federal government's acknowledgement of this responsibility.

Similar to Case 2, this NBS case involves unique privacy and information protection risks raised by repurposing data: moving data that was originally collected for a specific purpose in one domain (State public health) to another domain (Common Rule research institution) in which it will be analyzed for unrelated purposes (research). In addition, the evaluation of this NBS case must include both ethical horizon and ethical provenance perspectives because both researchers within the screening program, and those outside of the program who wish to use the materials from the program for their own research projects, have the responsibility to not damage or detract from the program in any way—i.e. as was seen in the Texas and Minnesota lawsuits. This heightened responsibility does not necessarily rule out repurposing the blood spots, or related information, for research outside the specific scope of the screening program, but it does increase the level of scrutiny and care that must be integrated into any such research program. Though a usual justification for pursing research that might include risks to research participants is the later, societal good that may come from the research, in the case of repurposing newborn screening materials and information one is weighing this possible future benefit of the research against potential harms to a program with well-established societal benefit. In addition, the societal contract undergirding the newborn screening program involves de facto mandatory participation. Involvement of the individuals whose blood spots and health information might then be used for research outside the scope of the program does not have this same social mandate, even if a particular State's guidelines for the screening program include a provision for the possible use of these materials in future research projects outside the scope of the program itself. This difference in the social mandate for research that repurposes the newborn blood spots is grounded, in part, in the difference in the social benefit of the screening program that has already been widely experienced.

In light of the above argument regarding the need for increased research participant protections for Newborn Screening Program data, and the elimination of the informed consent waiver by the Reauthorization Act, significant ethical concerns are raised by the procedures that appear to be proposed by the recent NPRM for research with Newborn Screening Program materials.

> As is stated in the NPRM: The proposed exemption category at §__.104(f)(2) requires that the privacy safeguards at§__.105 are met, and that broad consent to the earlier storage or maintenance of the biospecimens and information had already been obtained consistent with the requirements of §__.104(f)(1). This means that for secondary research using biospecimens informed consent must have been obtained using a consent form using the Secretary's template. **It is presumed that research involving newborn blood spots would frequently take place using this provision.** (Federal Register 2015: 53967)

Considering the fact that the NPRM does not clarify exactly how its proposed broad consent process would differ, or improve upon, the "opt out" consent approach currently used by the States, we cannot clearly compare and contrast our proposed DMP approach to that referenced in the NPRM. However, two main issues arise from this NBS case that any informed consent approach needs to address. First, since many, if not most, of the proposals for research on NBS materials will include blood spots that are already in the State biorepositories, any proposed consent process and DMP will have to address both blood spots that have been collected (questions of ethical provenance) as well as those to be collected in the future (questions of ethical horizon). The DMP decision tree walks investigators through the same rigorous process of assuring research participant protections for obtaining new biospecimens as for repurposing existing newborn blood spot repositories.

Second, as mentioned above, the NBS program has provided obvious benefit to society across our country, with little public controversy or complaint with regard to the goals of the program. If repurposed research is to be done on the samples and information collected by these State programs, it must be done in a way that does not generate public controversy or concern that may detract from or diminish the social benefits of the NBS program. It remains an open question as to whether or not any broad consent process for secondary research with NBS materials can successfully address the concerns of the public that might be raised with regard to privacy and information protections. In light of this issue, we argue that a much more valid and viable approach is to use our DMP for any outside research project along with extensive public engagement to get feedback regarding what research is judged to be worth the risk to the NBS program and what is not. Without this level of rigorous research evaluation, the risk of eroding the public trust in a program as beneficial as the NBS program is not worth taking. Scientists too often ignore that risk, approaching the public with "a voice that says, 'We have a consensus; now accept it,'" rather than giving the public an opportunity to participate in the process (Jamieson 2015, "Communicating the Value and Values of Science").

6.4 Case 4: Reusing Existing Data for New Research

Data Management Plan

Report Form

For use with the DMP Decision Tree

Scientists at Arizona State University conducted a series of investigations on blood samples obtained from the Havasupai Indians, a small tribe of people living at the bottom of the Grand Canyon. The studies began when the Havasupai approached ASU for help in understanding the high prevalence of diabetes among their people. In addition to conducting research on the possible existence of a genetic basis for diabetes as among the Pima, ASU scientists re-used blood samples drawn from individual members of the Havasupai tribe to conduct and publish results on multiple studies of which the Havasupai had no knowledge on a range of other disorders and characteristics. Upon accidentally hearing of the secondary use of the blood samples, the Havasupai sued ASU who, after paying over $1.7 M in legal fees, settled the case by paying $700,000 in reparations and providing other benefits in kind (Harmon 2010; Jacobs et al. 2010). This case marks an example of unauthorized reuse of previously collected biospecimens. Please note: we have imagined how we would have answered the questions in the decision tree for this project had we been the investigators and had the decision tree been available. All text is hypothetical but plausible for the purposes of illustrating the DMP decision tree.

Section 1: Describe the Data

A. Will the data be collected as responses/observations from individuals?

   a. Yes
   b. No: the project will employ previously collected biospecimens (blood)

B. Will the data be curated (taken from existing sources) from domains governed by the Common Rule?

   a. Yes: The biospecimens proposed for use in this project were originally collected as part of a project by investigators from our same university but for a different purpose. The original investigators collected blood from the Havasupai Indians to determine if they had a genetic predisposition for diabetes. The leadership of the Havasupai agreed to the study. All participants signed informed consent forms. The university IRB reviewed and approved the protocol for the study.
   b. No

C. Are you recombining various data from a variety of sources?

   a. Yes
   b. No: all data comes from the original Havasupai study.

D. Will new information be derived from the data?

   a. Yes: we will be using the blood samples for a different purpose than the original study, namely to study the genetic predisposition of the Havasupai for schizophrenia.
   b. No:

E. If none of these collection/curation approaches pertains, please describe the origins of the data you will use in the project.

F. Describe how the data to be collected, curated or derived will be FIPPS compliant.

   a. Notice: In order to use these data for a purpose other than originally stipulated and granted consent, we will contact the Havasupai leadership again to explain the new project, seek their consent and inform participants in the original study.
   b. Choice: we will offer the original participants the option to opt out of this study even though they agreed to the original study.
   c. Onward Transfer: the IRB of the university will review our notice and consent procedures and monitor their implementation to ensure proper transfer of authorized blood samples to our study.
   d. Security: we have consulted with the university's custodians of the blood samples and have documented that they are taking proper precautions to protect their physical security. We have also consulted with the university's computer security staff to ensure that the computerized files containing the

personally identifiable results of the analyses of the biospecimens are subject to information protections sufficient to the threats and consequences of unauthorized disclosure.

e. Data integrity: We will not use the Havasupai biospecimens for this study unless or without authorization from the Havasupai leadership and individual participants because it entails using the information for a different research topic than originally collected.
f. Access: We will inform the Havasupai leadership and participants in the study that they may gain access to, and, when necessary, correct any information about participants in the study.
g. Enforcement: the Arizona State University has developed procedures through its Office of University Counsel, Office of Compliance and Ethics and its Institutional Review Board to ensure compliance with the principles of fair information practice.

G. Briefly describe how the budget supports the collection activities.

a. The budget includes support for contacting the Havasupai leadership and soliciting consent for use of the biospecimens in this study from all participants.
b. The budget includes support for investigators to retrieve the biospecimens from storage, conduct new investigations, complete relevant analysis and prepare reports for presentation and publication.

Section 2: Describe the data "storing"

E. Will you possess/collect/curate de-identified data?

c. Yes
d. No: All data in this study will be personally identifiable biospecimens.

D. Describe your plans for protecting the

a. confidentiality of the data: We have consulted with the university's computer security staff to ensure that the computerized files containing the personally identifiable results of the analyses of the biospecimens are subject to information protections sufficient to the threats and consequences of unauthorized disclosure (see the Arizona State University Information Security Policy with special reference to personally identifiable information).
b. security of the data: We have consulted with the university's custodians of the blood samples and have documented that they are taking proper precautions to protect their physical security (see Arizona State University Policies and Procedures for Protecting Biospecimens).
c. varying access permission levels: other than authorized university biospecimen custodial staff, only authorized investigators and research technicians on this project will have access either to the biospecimens or the analytic results of this study.

E. Briefly describe your plan for working with, and (when necessary) training the information technology staff of your own and partnering institutions to assure implementation of your privacy, confidentiality, and security safeguards.

   a. We have consulted with the university's computer security staff to ensure that the computerized files containing the personally identifiable results of the analyses of the biospecimens are subject to information protections sufficient to the threats and consequences of unauthorized disclosure (see the Arizona State University Information Security Policy with special reference to personally identifiable information).
   b. Our consultations have included informing the university's computer staff and the members of our research team of the sensitivity of both the genetic information being derived from the biospecimens and of the vulnerability of the population under study.
   c. We will use data from no other research institutions in this study.

F. Briefly describe how the budget supports the data storage/management activities, including activities beyond the funding period.

   a. The budget includes cost recovery fees for storage and maintenance of the biospecimens during the research project. The university supports the biospecimen storage unit as a core facility of its ongoing research program.
   b. Indirect cost recovery supports the university's core computer facility and staff.

Section 3: Describe the data access plans

A. Will access to the data be shared either fully or partially during the project?

   a. Yes:
   b. No: Other than authorized university biospecimen custodial staff, only authorized investigators on this project will have access to the biospecimens or the personally identifiable analytic results of the study. Only Arizona State University investigators will work on this project.

B. Will access to the data be shared either fully or partially after the funding ends?

   a. Yes: the Havasupai biospecimen collection is available to authorized investigators who comply with the procedures for obtaining consent from the Havasupai leadership and original participants and for protecting the biospecimens and personally identifiable results.
   b. No

C. What are the possible harms to your research subjects of full or partial sharing?

   a. If not shared with proper consent or appropriate data protection measures, multiple harms could accrue to the Havasupai, the Arizona State University and genomic science, including:

i. Dishonor and embarrassment to individual Havasupai people with the breach of the confidentiality of their personally identifiable data;
ii. Secondary reuse of the Havasupai biospecimens without consulting and receiving approval of the Havasupai leadership affronts the community's sovereignty.
iii. Arizona State University could suffer public embarrassment, financial penalties, needless legal expense, wasted time of investigators spent defending their actions rather than doing science, and disruption of their relationship with the Havasupai.
iv. The institutions of science, particularly genomic science, could lose the trust of the public including but not limited to indigenous peoples like the Havasupai who interpret scientific behavior in the context of their colonial experiences.

D. Explain permission or any limitations that may exist on the use or reuse of the data in your submission to the IRB?

a. We have consulted with the university IRB throughout planning of this project.
b. We have submitted and received approval for our human subjects review packet with informed consent form.
c. We intend to keep the IRB informed of our progress in negotiations with the IRB, including providing it with a list of subjects who opt out of having their biospecimens included in this study.

E. Describe the process by which a data dictionary will be created and maintained
F. Briefly describe how the budget supports the data access plans: The budget includes no funds for data access because no investigators from outside the university will participate.
G. Briefly describe the timeline for implementing data access plans. Not applicable

Commentary on Case 4: Decision Tree in Case of Unauthorized Reuse

This hypothetical example of completing the DMP decision tree before beginning the reuse of the original Havasupai data offers important lessons. First, with respect to the ethical provenance of the Havasupai data, the University of New Mexico thoroughly established the legitimacy and acceptance of the original project by working closely with the Havasupai leadership, gaining approval from the university's IRB, and obtaining informed consent from potential participants. Subsequent researchers could have capitalized upon and extended this legitimacy by returning to the Havasupai and requesting permission to conduct the later experiments. As illustrated by the hypothetical example here, the DMP decision tree should have led them to adopt that approach specifically by prompting them on the consent issue. Two parts of the decision tree seem highly relevant, the question associated with FIPPS compliance and the questions associated with potential harms. The FIPPS compliance questions prompt investigators to query what original participants intended when consenting to the initial study and requires investigators to give them a chance to refuse participation in a subsequent study. The FIPPS questions also draw attention to information protection issues and, thus, encourage investigators to consult with their organizational computer or information security staff. The harms question invites incorporation of the privacy matrix we have explained elsewhere (Steinmann et al. 2015) because it outlines the range of ethical implications possible in any given case. Note, too, that various university "compliance" agencies have roles in managing this hypothetical case, including but not limited to the IRB.

The NPRM initially takes a strong stance on reuse of biospecimens by saying

> Regardless of the scale on which harms may have occurred in the past, continuing to allow secondary research with biospecimens collected without consent for research places the publicly funded research enterprise in an increasingly untenable position because it is not consistent with the majority of the public's wishes, which reflect legitimate autonomy interests. (Federal Register 2015: 53944).

The NPRM would also certainly treat this study as an example of human subjects research and, indeed, cites the Havasupai case (Federal Register 2015: 53943). Yet, it softens its stance later in the document when considering the question of the exemption of secondary research on biospecimens or identifiable private information. For cases just such as this in which "biospecimens were or will be collected for a different research study or for a non-research purposes" (Federal Register 2015:53972), the NPRM proposes to allow exempt review in which the "reviewing IRB conducts a limited IRB review of the process through which broad consent will be sought, and, in some cases, of the adequacy of the privacy safeguards described in §__.105." (Federal Register 2015: 53966). At the time of this writing, the NPRM had not given many details about design or implementation of the concept of "broad concept". Its skeletal outline suggests, however, that it would not suffice for the Havasupai case, primarily because it would not meet the community's expectations about individual consent and community sovereignty. Given, too, the list of potential and, indeed, actual harms from authorized secondary reuse of the Havasupai biospecimens, this type of case warrants full board review and thorough consent not exempt or limited IRB review using a broad consent mechanism however it may ultimately be defined.

6.5 Case 5: Reanalysis of Data for Unspecified Future Research

Data Management Plan

Report Form

For use with the DMP Decision Tree

The Million Veteran Program (MVP), a Department of Veterans' Affairs (VA) genomic biobank initiated and conducted under the leadership of one of the coauthors (JK), will link phenotypes derived from the VA Electronic Health Record

(EHR) and other sources to genetic and exome data. It is part of the Cooperative Studies Program, the major clinical trials entity of VA research and is now also part of the President's Precision Medicine Initiative. As of September 2015 about 415,000 veterans had enrolled with plans to recruit 1 million veterans by 2018. Research projects will include the genetic basis of veterans' diseases (starting with mental health, e.g. vulnerability and resilience to PTSD), diagnostic testing, targets for drug development, pharmacogenetics, Point of Care Research, and others. The program entails close collaboration of researchers with the VA healthcare system and included a selection process for VA Medical Centers wishing to be MVP sites.

Section 1: Describe the Data

A. Will the data be collected as responses/observations from individuals?

   a. Yes: The MVP is forming a database which includes access to Electronic Health Records (EHR) and a collection of biospecimens for as-of-yet unspecified research projects not a research project per se. All veterans in VA healthcare are eligible for recruitment. The enrollment process starts with letters of invitation. Veterans then have the option to decline participation and further contact or if they wish to participate, they complete a baseline survey and, at the next VA visit, sign an informed consent and HIPAA authorization for future examination of their EHR and allow secure access to VA and VA-linked medical and health information. Veterans also provide a blood sample (serum and Buffy coat). A Certificate of Confidentiality, according to National Institute of Health guidelines, is supplied. Veterans' samples and health samples and data are labeled with a code stored in a secure biorepository and database behind a VA firewall. Future studies using these data require further consent (except in certain very specific instances) and IRB review and approval on a project-by-project basis.
   b. No

B. Will the data be curated (taken from existing sources) from domains governed by the Common Rule?

   a. Yes: Existing sources include EHR information from the Veterans Health Administration, such as history, physical exam, laboratory tests, etc. The veterans who volunteer for the MVP understand that they are contributing specimens to a collection for use in as-of-yet unspecified research projects by future authorized but not yet identified investigators. The informed consent form includes a full explanation of these conditions.
   b. No

C. Are you recombining various data from a variety of sources?

   a. Yes: Future research projects may encompass a variety of sources, including Medicare, Department of Defense, academic medical center and other private sector data as well as the MVP biospecimen collection. A variety of

measures help protect the privacy of veterans and the security of their information in the MVP, including limiting access to authorized researchers affiliated with the VA and a full range of computer security policies, procedures and practices measures (see below for further details). Processes in MVP, including regarding security, are communicated on websites (http://www.research.va.gov/for_veterans/safeguarding_vets.cfm).
   b. No

D. Will new information be derived from the data?

   a. Yes: Because the research projects remain to be defined, a precise answer to this question does not exist. It is reasonable to assume that some will derive new information from the underlying data.
   b. No

E. If none of these collection/curation approaches pertains, please describe the origins of the data you will use in the project.
   N/A
F. Describe how the data to be collected, curated or derived will be FIPPS compliant.
   Notice: As investigators propose projects, they must explain and obtain IRB approval for access to, and use of data in the MVP
   Choice/Consent: When veterans volunteer for the MVP, they understand that their biospecimens may be used in future as-of-yet undefined projects. They also understand that the VA IRB will review and approve such uses in the future. For all future projects, MVP veteran volunteers will have the opportunity to grant or withhold informed consent.
   Access: Veterans who volunteer for the MVP are informed they may gain access to, and, when necessary, correct any information about themselves in the study.
   Security: the VA takes stringent measures to protect the confidentiality and security of the data, including rigorous technical defenses and implementation of all federal computer security protection laws, regulations and standards.
   Enforcement: the VA has developed procedures through its Office of General Counsel, Office of Research Oversight, the Office of Information and Technology, and Institutional Review Board to ensure compliance with the principles of fair information practice. In addition, strong oversight of the program exists under the VA Cooperative Studies Program, including unannounced audits
G. Briefly describe how the budget supports the collection activities.
   The budget for MVP falls within the federal budget process for VA, which is an annual appropriation. Funding for IT in the VA is sequestered with the IT budget administered by the VA's Office of Information and Technology but specific research IT programs may be funded within the Office of Research and Development. In addition, for individual projects, investigators may obtain research funding from other sources, including NIH, foundations, and other federal departments.

Section 2: Describe the data "storing"

A. Will you possess/collect/curate de-identified data?

   a. Yes: Individual data will be provided to investigators with name, Social Security number, date of birth and address omitted (as in the UK Biobank).
   b. No

B. Describe your plans for protecting the

   a. confidentiality of the data: Veterans' samples and health samples and data are labeled with a code stored in a secure biorepository and database behind a VA firewall. Only a small group of authorized VA personnel will have access to the code. Individual data will be provided to investigators with name, Social Security number, date of birth and address omitted (as in the UK Biobank).
   b. security of the data: the VA takes stringent measures to protect the confidentiality and security of the data, including rigorous technical defenses and implementation of all federal computer security protection laws, regulations and standards.
   c. varying access permission levels: Only authorized investigators affiliated with the VA and working under the auspices of an approved project may have access to the MVP database.

C. Briefly describe your plan for working with, and (when necessary) training the information technology staff of your own and partnering institutions to assure implementation of your privacy, confidentiality, and security safeguards.

   There is an Honest Broker approach, vetting process and required Data Use Agreements for data users. As indicated above, all Information Technology in the VA is under the Office of Information and Technology (OIT). OIT administers the IT compliance program in cooperation with the Office of Research and Development, which oversees VA research.

D. Briefly describe how the budget supports the data storage/management activities, including activities beyond the funding period.

   The budget for MVP falls within the federal budget process for VA, which is an annual appropriation. Funding for IT in the VA is sequestered with the IT budget administered by the VA's Office of Information and Technology but specific research IT programs may be funded within the Office of Research and Development. In addition, for individual projects, investigators may obtain research funding from other sources, including NIH, foundations, and other federal departments.

Section 3. Describe the data access plans

A. Will access to the data be shared either fully or partially during the project? Sharing of data from the MVP occurs on a project-by-project basis and requires peer-reviewed and IRB approval of the protocol. Data sharing was the subject of considerable discussion within VA and with other agencies. Principle investigators must be VA employees. Partnering on specific projects is one approach to sharing data with investigators who do not work for the VA because co-investigators do not have to be VA employees.

B. Will access to the data be shared either fully or partially after the funding ends?

   Sharing of data from the MVP occurs on a project-by-project basis and requires peer-reviewed and IRB approval of the protocol.

C. What are the possible harms to your research subjects of full or partial sharing?

   Since MVP data are extensive and include all individual health and genomic information, much harm could come from loss of privacy. Such information could be stigmatizing to individuals and could be used for denial of insurance, inappropriate marketing, identity theft, and other specific harms. Group harm could also come to veterans from certain revelations. While government databases have firewalls, they may be subject to Freedom of Information Act requests with certain restrictions including de-identification.

D. Explain permission or any limitations that may exist on the use or reuse of the data in your submission to the IRB.

   a. Yes: All such information must be submitted to an IRB on future projects.
   b. No

E. Describe the process by which a data dictionary will be created and maintained
   The computing infrastructure for MVP includes the VA Informatics and Computing Infrastructure (VINCI) which hosts variety of VHA databases with data from all veterans, software analysis and reporting tools for all projects in a high-performance analytic environment. The VA Consortium for Healthcare Informatics Research (CHIR) offers for natural language processing. The VA Genomic Information System for Integrative Science (GenISIS) provides genomic support and infrastructure including management of recruitment and enrollment, blood sample tracking, genomic datasets and the scientific environment for analysis. All of these IT environments are behind firewalls.
F. Briefly describe how the budget supports the data access plans
   The budget for MVP falls within the federal budget process for VA, which is an annual appropriation. Funding for IT in the VA is sequestered with the IT budget administered by the VA's Office of Information and Technology but specific research IT programs may be funded within the Office of Research and Development. In addition, for individual projects, investigators may obtain research funding from other sources, including NIH, foundations, and other federal departments.
G. Briefly describe the timeline for implementing data access plans.
   Because access to the MVP data occurs on a project-by-project basis, no specific timeline exists for data sharing.

Commentary on Case 5: Reanalysis of Data Collected for Use in As-of-yet Not Proposed Research Projects

This case illustrates the importance of the concept of ethical horizon and future-oriented consent when analyzing human subjects research with big data. The MVP collects from veterans biospecimens and data from other databases for use in future research projects in which neither the investigator, the purpose, the exact data requirements nor the risks are known. The NPRM does not precisely identify this type of situation because the MVP collects biospecimens for future unspecified uses without ever being included in an original study. Under section B.2.d. it does say,

> the NPRM proposes to allow broad consent to cover the storage or maintenance for secondary research use of biospecimens and identifiable private information. Broad consent would be permissible for the storage or maintenance for secondary research of such information and biospecimens that were originally collected for either research studies other than the proposed research or non-research purposes. The broad consent document would also meet the consent requirement for the use of such stored biospecimens and information for individual research studies (Federal Register 2015: 53972–73).

If we propose that being collected for some future unspecified use compares closely with the status of having been collected "for research studies other than the proposed research", we may surmise that the NPRM would favor employing a broad consent approach to obtaining consent for the MVP. At first glance, the MVP appears to employ a kind of broad consent because veterans agree to contributing their biospecimens and other identifiable information without knowing what research might eventually be conducted with their information. Yet, the VA also chose to solicit additional consent from veterans whose information gets selected for use in an actual study later. Why did the VA take this step?

When creating the program, the VA emphasized the importance of obtaining and sustaining the trust of the veterans for success in obtaining, using, analyzing and applying data for the long-term benefit of veterans. The VA took multiple steps to engender trust in MVP, including:

- Commissioned a survey of veterans (with preceding focus groups) by a non-VA group, the Genetics and Public Policy Center. The survey found strong support for the MVP (83 % approved of it and 71 % said they would consent), support for sharing data with academic institutions and government, and very strong concern about privacy—98 % both wanted safeguards to protect information from misuse and disclosure and serious consequences for researchers who violate the research agreement. Over 90 % also wanted it to be illegal for insurers, employers to get the information, 87 % for law enforcement (Kaufman et al. 2009)
- Communicates on websites the processes in MVP, including regarding security (http://www.research.va.gov/for_veterans/safeguarding_vets.cfm).
- Sponsor central IRB and peer review for all requests for access to the database

- Implement an Honest Broker approach and vetting process for data users with required Data Use Agreements (Auray-Blais and Patenaude 2006; Dhir et al. 2008; Taube et al. 1998)
- Solicit initial consent only for sample collection and access to data with further consents needed for research projects
- Rigorous information protection policies, procedures, and practices
- The VA Secretary, Deputy Secretary and Chief of Staff enrolled in the first group

The VA approach, which was put into place prior to the NPRM, has effectively implemented a strong version of the NPRM's observation that the American public expects to give consent for secondary use of biospecimens in order to engender trust in the institution of science, though it is a different situation than future use of a consented study. With this case, we return full circle to issues raised initially by the Havasupai. What lessons should we draw from these cases and these differing approaches to reusing and prospective reanalysis of biospecimens? The Havasupai and veterans are identifiable as groups and we could classify both as populations whose relationship to the rest of American society puts them at special risk for harm in case of breaches of their privacy or autonomy (an "exceptionalist" approach). This approach might argue that broad consent (however it becomes defined) for secondary reuse of biospecimens suffices for the general population but, depending on the circumstances, requires modification when applied to vulnerable or otherwise at risk populations, although in the case of veterans, future use of biospecimens is different from the situation of an initial consented study. The MVP is a biobank and not a study per se. An "inclusionist" approach might argue, to the contrary, that the similarities between the Havasupai, veterans, and the rest of American society with respect to the risks of misusing information derived from biospecimens warrant similar approaches to both consent and data stewardship. An inclusionist approach might suggest that the utility of the concept of "broad consent" lies in alerting human subjects of the ethical horizon of their participation in an initial study; that is, informing them that, in the contemporary world of biomedical science, a strong possibility exists that their samples could, and, probably, would be selected for study in future, as-of-yet unspecified research projects by to-be-identified investigators. The inclusionist approach, however, also implies that broad consent should not obviate the need for obtaining secondary informed consent from human subjects whose samples later scientists select for study.

7 Conclusion

Given the uncertainties of big data research about human subjects as illustrated in the cases of this chapter, we suggest that investigators should engage in privacy and ethical analysis of their proposed data and projected results as they are preparing their initial proposals for funding as well as during and after they conduct their research. We offer the decision tree for preparing the Data Management Plan as a tool for aiding in these analyses. The decision tree encourages investigators to reflect on the ethical status of any consent provided by the human subjects under consideration, if known or potentially known, for both existing data and data proposed for collection. The decision tree also encourages investigators to reflect on the range of ethical implications of their research, including potential benefits and harms as well as implications for individual and group autonomy, social justice and trust in the institution of science. We observe that such ethical implications may arise across the full trajectory of a research project, including from before its inception when using information obtained from existing databases and after its completion when projecting secondary use of data by subsequent investigators. The decision tree encourages investigators to reflect on these issues when making plans for sharing data. The decision tree also encourages investigators to consult with key stakeholders in their own institutions and other institutions that might potentially grant access to data under their control. With respect to big data projects, the decision tree prompts investigators to hold these consultations particularly with the Institutional Review Board, the information and computer security staff of participating institutions, and, when necessary, general counsel. For topics engaging complex combinations of data types from multiple, disparate sources, we recommend incorporating experts in information security and ethical analysis into the research team.

We adopted this approach to the DMP decision tree because our experience suggests the scientific community, in general, has insufficient experience working with big data in all its various forms to waive traditional human subjects protection such as consent. In addition, some segments of the scientific community now taking an interest in big data research on human subjects have almost no experience with institutional review boards or the tradition of ethical reflection and scientific controversy underpinning them. Thus, we take exception to the attempt of the NPRM to downgrade reflection on, and review of big data research on human subjects, particularly secondary analysis of biospecimens and publicly available documentary, supposedly de-identified or anonymous data sets. We think the NPRM's analysis fails to address the privacy and ethical implications of big data research with human subjects across the full spectrum of research. For example, when suggesting that publicly available data pose no new risks of harm to human subjects just because they are already public, the NPRM ignores the possibility of new information arising in the course of initial analysis, reuse, repurposing, recombination, or reanalysis. As Gymrek et al. (2013) demonstrated, it is possible to identify individuals by combining surnames with data obtained from free, publicly accessible Internet resources, such as age and state. Surnames were recovered from personal genomes using Y-chromosome short tandem repeats and recreational genetic genealogy databases (Gymrek et al. 2013: 321–324; see also Rodriguez et al. 2013). It bears notice that, in some cases, the complexity of big data as proposed in social science projects derives precisely from the difficulty of employing traditional human subjects protections such as consent while, nonetheless, potentially putting human subjects at risk of harm. Thus, through the DMP decision tree described in this chapter, we offer a tool to help investigators become better aware of the privacy and ethical implications of their work across the full spectrum of acquiring, analyzing, sharing and taking action on the basis of big data.

Finally, the themes of our analysis suggest stronger collaboration among investigators using big data for human subjects research and various departments in their own and partner's organizations such as the IRB and university information services. The NPRM casts the relationship between investigators and such departments (especially IRBs) as burdensome, relatively unproductive and overwhelmed by paperwork. The suggestion also exists that the true hazards of research (including big data research) focus on biomedical sciences rather than research in the social, behavioral, or economic sciences. If by "big data" we simply mean the so-called 3 Vs of volume, velocity and variety, the suggestion might have merit. In this chapter, however, we have emphasized the 4 Rs of reuse, repurposing, recombining, and reanalysis. The 4 Rs entail questions about the ethical provenance and ethical horizon of big data not just its quantity, speed or heterogeneity. The 4 Rs also draw attention to a range of potential harms and unacceptable ethical consequences of human subject research with big data that the NPRM disregards. We think that few investigators will have the ethics or computer security training to understand all these possibilities at the beginning of a project. With the aid of experts in these matters during the planning, execution and evaluation of big data projects, they can and will gain a better understanding of potential risks to the data and the human subjects. As investigators and their home organizations gain this experience, they may gradually learn to distinguish between more or less risky projects employing big data in human subjects research and to adopt appropriate approaches to their review and management. The NPRM seems to suggest that investigators and Common Rule research organizations own that experience now. With all due respect, we disagree. We also offer the DMP decision tree as a tool to help investigators and their organizations become more adept in assessing the privacy and ethical risks of big data research with human subjects and, thus, ensure the public's acceptance and participation in the projects they plan for the future.

References

Auray-Blais, C., & Patenaude, J. (2006). A biobank management model applicable to biomedical research. *BMC Medical Ethics, 7*(4). doi:10.1186/1472-6939-7-4

Berkowitz, S., et al. (2014, July 21). *Organization of static and dynamic ontological concepts that describe forced migration from and displacement within Somalia.* Presented at the Humanitarian innovation Conference, Oxford, UK.

Collmann, J., & Robinson, A. (2010). Designing ethical practice in biosurveillance: The project argus doctrine. In D. Zeng, H. Chen, C. Castillo-Chavez, B. Lober & M. Thurmond (Eds.), *Infectious disease informatics and biosurveillance: Research, systems, and case studies* (pp 24–44). New York: Springer.

Collmann, J., et al. (2014a, July 21). *Menacing context links forced migration drivers to family decision-making.* Presented at the Humanitarian innovation Conference, Oxford, UK (submitted for publication).

Collmann, J., Berkowitz, S., & Singh, L. (2014b). Using large-scale open source data to identify potential forced migration. data science for the social good—ACM Knowledge Discovery and Data Mining (KDD) Workshop.

Dhir, R., Patel, A. A., Winters, S., Bisceglia, M., Swanson, D., Aamodt, R., et al. (2008). A multidisciplinary approach to honest broker services for tissue banks and clinical data. *Cancer, 113*(7), 1705–1715. doi:10.1002/cncr.23768.PMC2745185.PMID18683217

Federal Policy for the Protection of Human Subjects; Notice of proposed rulemaking. 80 Federal Register 173 (2015, September 8), pp. 53933–54061.

FTC Fair Information Practice. In Wikipedia. Retrieved November 2015, from https://en.wikipedia.org/wiki/FTC_Fair_Information_Practice

Gymrek, M., McGuire, A. L., Golan, D., Halperin, E., & Erlich, Y. (2013, January 18). Identifying personal genomes by surname inference. *Science, 339*(6117), 321–324. doi:10.1126/science.1229566

Harmon, A. (2010, April 21). Indian tribe wins fight to limit research of its DNA. *NY Times*. http://www.nytimes.com/2010/04/22/us/22dna.html?pagewanted=all&_r=0. Accessed 5 October 2015.

Hirschhorn, L. (2014, July 21). *Building community in a transdisciplinary setting: The forced migration group as a developmental project.* Presented at the Humanitarian innovation Conference, Oxford, UK.

Jacobs, B., Roffenbender, J., & Collmann, J. (2010). Bridging the gap between genomic scientists and indigenous peoples. *Journal of Law, Medicine and Ethics, 38*(3), 684–696. http://arep.med.harvard.edu/pdf/Jacobs-JLME_10.pdf

Jamieson, K. H. (2015, October 7). *Communicating the value and values of science*. Presented at the Special CSAP Lecture Event, Yale Center for the Study of American Politics, New Haven, CT.

Kaufman, D., Murphy, J., Erby, L., Hudson, K., & Scott, J. (2009). Veterans' attitudes regarding a database for genomic research. *Genetics in Medicine, 11*(5), 329–37. doi:10.1097/GIM.0b013e31819994f8

President's Council of Advisors on Science and Technology, Big Data and Privacy: A Technological Perspective. Report to the President May 2014. https://www.whitehouse.gov/sites/default/files/microsites/ostp/PCAST/pcast_big_data_and_privacy_-_may_2014.pdf. Accessed 5 October 2015.

Rodriguez, L. L., Brooks, L. D., Greenberg, J. H., & Green, E. D. (2013). The complexities of genomic identifiability. *Science, 339*(6117), 275–276. doi:10.1126/science.1234593

Steinmann, M., Shuster, J., Collmann, J., Matei, S., Tractenberg, R., FitzGerald, K., et al. (2015). Embedding privacy and ethical values in big data technology. In S.A. Matei, M. Russell, E. Bertino (Eds.), *Transparency on social media—tools, methods and algorithms for mediating online interactions.* New York: Springer.

Taube, S. E., Barr, P., Livolsi, V., & Pinn, V. W. (1998). Ensuring the availability of specimens for research. *The Breast Journal, 4*(5), 391. doi:10.1046/j.1524-4741.1998.450391.x

Zimmer, M., & Proferes, N. J. (2014). A topology of twitter research: Disciplines, methods, and ethics. *Aslib Journal of Information Management, 66*(3), 250–261. doi:10.1108/AJIM-09-2013-0083. Accessed 6 November 2015.

Integrating Ethical Reasoning into Preparation for Participation to Work in/with Big Data Through the Stewardship Model

Rochelle E. Tractenberg

> Scholars who study the behavior of human organizations constantly stress the importance of defining institutional goals. Without clear objectives, it is said, an institution cannot evaluate how well it is performing, decide how to allocate its resources wisely, plan for future growth, motivate its members, or justify its existence to the larger public.
>
> Derek Bok, "On the purposes of Undergraduate Education" (1974), p. 159

1 The Stewardship Model and Ethical Reasoning

In 2001, the Carnegie Foundation for the Advancement and Scholarship of Teaching instituted a 5-year, in-depth review of doctoral training in the United States, the Carnegie Initiative on the Doctorate (CID). At the initiation of the CID project, the Foundation articulated that the purpose of doctoral education is "…to educate and prepare those to whom we can entrust the vigor, quality, and integrity of the field. This person is a scholar first and foremost, in the fullest sense of the term—someone who will creatively generate new knowledge, critically conserve valuable and useful ideas, and responsibly transform those understandings through writing, teaching, and application. We call such a person a "steward of the discipline."" (Golde 2006: p. 5)

The model of "disciplinary stewardship" was focal for the CID, and yet has not spawned much empirical (quantitative or qualitative) work beyond the Foundation's two edited volumes, "Envisioning the Future of Doctoral Education" (Golde and Walker 2006) and "The formation of Scholars: Rethinking doctoral education for the twenty-first century" (Walker et al. 2008). This might be due to the fact that the

---

R.E. Tractenberg (✉)
Collaborative for Research on Outcomes and -Metrics; and Departments of Neurology, Biostatistics, Bioinformatics and Biomathematics, and Rehabilitation Medicine, Georgetown University Medical Center, Building D, Suite 207, 4000 Reservoir Rd. NW, Washington, D.C., NW 20057, USA
e-mail: rochelle.tractenberg@gmail.com

J. Collmann and S.A. Matei (eds.), *Ethical Reasoning in Big Data*, Computational Social Sciences, DOI 10.1007/978-3-319-28422-4_11

primary fields within which the CID project and the construct of stewardship were explored during the project did not include Education; and that doctoral education is usually defined by considerations of the field in which the doctorate is pursued, rather than by considerations of Educational theory or research. This tends to limit the motivation for, say, neuroscience faculty to pursue scholarship in 'doctoral education'—because their stewardship is naturally focused primarily on *neuroscience*. On the other hand, the definition and ideals of disciplinary stewardship have been highly influential in my own professional identity development as well as in the curricular development and evaluation methodology I developed, the Mastery Rubric (described below). As a cognitive scientist and statistician who supports doctoral training programs by teaching stand-alone service courses in statistics, methods, and research ethics, I have dedicated a great deal of time to their roles in other programs. The consideration of how these "auxiliary" courses can promote—or at least contribute to the promotion of—the stewardship of all students' respective disciplines has been fascinating; it has also supported a portfolio of scholarship of teaching and learning. I have been fortunate enough to have had the opportunity to develop that portfolio while my stewardship of the disciplines for which I completed doctoral training has simultaneously been deepening.

This chapter is written from the perspective that disciplinary stewardship is actually a more general model for "education and preparation", and as such, its introduction need not be postponed until, or limited to, doctoral training. There are features of the steward of a discipline that can be initiated or introduced earlier than the doctoral level: three examples are that: (1) disciplines and fields are dynamic, and require stewardship; (2) the quality and integrity of disciplines must be actively preserved and conserved; and (3) there are particular habits of mind that characterize "those to whom we can entrust" the core features of a discipline or field. The concept that undergraduate and master's level education should prepare students for the workforce is not new (see, e.g. Bok 1974); what is suggested here is that integrating ethical reasoning into higher education generally requires "clear objectives", and the stewardship model can provide these. It must be noted that "integration of ethical reasoning into higher education" is not a clear objective, nor is "preparation to work in/with Big Data".

In order to formulate clear objectives that will support the integration of ethical reasoning across the university in an evaluable way, doctorally-trained faculty need to perceive the relevance of ethical reasoning to their own disciplinary stewardship. This is true for anthropologists, economists, mammalogists and zoologists, and nearly all disciplines in between, at any university—irrespective of whether they use big, small, simulated, or qualitative data in their work. Anthropology (e.g., American Anthropological Association http://www.aaanet.org/issues/policy-advocacy/upload/AAA-Ethics-Code-2009.pdf), mammalogy (e.g., American Society of Mammalogists http://www.mammalsociety.org/uploads/ASM%20Ethics%20Statement.pdf); and zoology (e.g., Zoological Association of America http://www.zaa.org/pdf/ZAA%20Code%20of%20Conduct%20-FINAL-%202.9.09-old.pdf) all have codes of ethics; economics may be getting one someday (DeMartino 2011). These codes represent professionalism and the habits of mind that all stewards of a discipline can buy into; we have argued that these codes can be harnessed to teach, give practice, and promote development in ethical reasoning that is discipline-specific (Tractenberg et al. 2014), and can seed ongoing growth in, and application of, these reasoning skills (Tractenberg et al., in review). Ethical reasoning, while not specifically described in any professional association code of conduct [although see the Association of Internet Researchers (AoIR) revised code of conduct, "Ethical Decisionmaking and Internet Research" (Association of Internet Researchers Ethics Committee 2012)], is nevertheless a widely recognized set of knowledge, skills and abilities: the identification and assessment of one's prerequisite knowledge; recognition of a moral issue; identification of relevant decision-making frameworks; identification and evaluation of alternative actions; making and justifying a decision (about the moral issue that was recognized); and reflection on the decision (Santa Clara University (no date), http://www.scu.edu/ethics/; see also Kligyte et al. 2008; Hollander and Arenberg 2009 for the same skills derived from non-ethics perspectives).

A clear and evaluable institutional objective that can support the integration of ethical reasoning across a university would be "one instructor in every degree-granting program will teach one required course in ethical reasoning for <their discipline>." The recognition that each department requires discipline-specific training in ethical reasoning represents one of the most basic aspects of stewardship: "disciplines and fields are dynamic, and *require stewardship*." The creation and active maintenance of the codes of ethical or professional conduct within each discipline instantiates the second principle of stewardship that can be introduced to students at all levels in that field: "the quality and integrity of disciplines must be actively preserved and conserved." And finally, the commitment by these departments and in particular, these instructors, to stewardship of their respective disciplines would represent the third element of stewardship that can be introduced to, and modeled for, students in a discipline prior to initiation of doctoral level training ("there are particular habits of mind that characterize 'those to whom we can entrust' the core features of a discipline or field.").

2 The Mastery Rubric and Integrating Ethical Reasoning into Higher Education

The Mastery Rubric is a tool for curriculum development and evaluation (Tractenberg 2013). A Mastery Rubric is created to describe the knowledge, skills and abilities that an entire curriculum is designed to deliver, and it also includes specific, concrete, but flexible descriptions of the performance of these knowledge, skills and abilities (KSAs) along a developmental trajectory as the learner changes—via the curriculum—from more novice performance and habits of mind towards those that are more similar to experts in the field. Even if an individual develops a single course, knowing where in the developmental trajectory the learner needs to be in order to contribute to/engage with/benefit from a course is useful (Tractenberg et al. 2010). However, the trajectory is critical if further development in those KSAs is ultimately to be sought—either opportunities to do so are sought by the students or opportunities to support it are desired by the institution (Tractenberg and FitzGerald 2012).

The first Mastery Rubric was published in 2010, and since that time this construct has been used to frame a variety of programs in training and higher education generally within this *developmental* perspective. The second Mastery Rubric to be published described a developmental approach to training in the responsible conduct of research through a focus on ethical reasoning (Tractenberg and FitzGerald 2012; see also Tractenberg 2013; Tractenberg et al. 2014; Tractenberg and FitzGerald 2015; Tractenberg 2016b). The third Mastery Rubric (Tractenberg et al., in press) describes a developmental approach to evidence-based medicine. This developmental approach to curriculum design in higher education meshes well with the CID stewardship model; if the characteristic KSAs required for stewardship of a discipline are identified, and the levels of performance on each KSA that characterizes the "steward" are articulated, then principles of backward design (Wiggins and McTighe 1998; see also Messick 1994) or constructive alignment (Biggs and Tang 2007) can be harnessed to integrate training in ethical reasoning into undergraduate, graduate and post-graduate/professional training (see, e.g., Tractenberg, this volume).

Irrespective of prior training and education, the preparation of individuals to work in and with Big Data requires some consideration of not just what ELSI might be encountered, but how to prepare these individuals to reason through or about these ELSI—and ones that are not yet envisioned. This can be facilitated by consideration of how ethical reasoning can help these individuals to appreciate that those *fields* wherein Big Data is used and/or generated *require stewardship* and that the *quality and integrity of these fields*—as well as this Big Data—*must be actively preserved and conserved.* By combining the stewardship model together with the Mastery Rubric approach to teaching ethical reasoning, departments and programs can emphasize for all students that there are particular habits of mind that characterize "those to whom we can entrust" the core features of a discipline or field and these include considerations around the ethical, legal, and social implications of Big Data (see, e.g., Trede 2012). Currently, a focus is on identifying single issues that might result in ethical challenges, and preparing training content targeting these specific issues. We have argued previously that the ability to effectively ethically reason through challenges is not a skillset that introduction to/case analyses of a single specific challenge will support (Tractenberg and FitzGerald 2012; Tractenberg and FitzGerald 2015). Again, we advocate inculcating participants in work involving Big Data with the idea that their professional identity embodies some responsibility to maintain this particular skillset with respect to known and future ELSI (Tractenberg et al. 2014; Tractenberg 2016b).

While the definition of disciplinary stewardship from the CID has been very influential in my own professional identity development, there is a related definition of "steward" that is specifically relevant for preparation for participation in/with Big Data, from the Gates Foundation (Bill and Melinda Gates Foundation 2011; p. 2): "All who produce, share, and use data are *stewards of those data*. They share responsibility for ensuring that data are collected, accessed, and used in appropriate ways, consistent with applicable laws, regulations, and international standards of ethical research conduct." (emphasis added). It is well known that participation in and with Big Data is not limited to those with Ph.Ds and moreover, that because "Big Data" is so amorphous and there are so many different paths that can lead to working with and in Big data, it might be difficult to ever identify those who are actually being "prepared to engage with and in Big Data" in the first place. Therefore, the approach that is outlined here is consistent with initiating *all* learners in the stewardship model (whether or not they continue on, to become actual stewards themselves); introducing ethical reasoning in *any* undergraduate or graduate program, and perhaps integrating a respect for ethical data practice into all engagement with data (Big, big, or small), could be important for improving the reproducibility of science across disciplines (see e.g. Freedman 2010; Collins and Tabak 2014; McNutt 2014).

3 A New Conceptualization for "Ethics Education"

Many—but not all—science PhD programs require a single ethics course (see Tractenberg 2016b; although see also Lee et al. 2015). The National Institutes of Health (NIH) in the United States, for example, requires that those being trained in research using NIH funds must complete training in the "responsible conduct of research" at least once in four years (NIH 2009). Although this policy also includes a comment that "Active involvement in the issues of responsible conduct of research should occur throughout a scientist's career" (http://grants.nih.gov/grants/guide/notice-files/NOT-OD-10-019.html; Basic Principles"), the policy is also *explicitly only* required for individuals whose research is specific to "human subjects" (see Tractenberg 2016b).

The clear and evaluable institutional objective articulated earlier ("one instructor in every degree-granting program will teach one required course in ethical reasoning for <their discipline>") implies that one course is sufficient to achieve the critical—and complex—set of ethical reasoning KSAs that encompasses "the ways in which research…(is) conducted in that particular discipline" (Shulman 2008). However, neither one institution-wide (generic/discipline-independent) "ethics course", nor encouraging students to learn what ethics they need as they go, can achieve this objective. Even a single discipline-specific course must be followed by additional opportunities to engage in, develop, and refine the skill set (of ethical reasoning). Having articulated the KSAs in ethical reasoning, and particularly how development in their performance would look [i.e., with a Mastery Rubric or specifically, the Mastery Rubric for Ethical Reasoning (2012)], the institution can (should) create new opportunities—and evaluate all training opportunities for their alignment with, and potential to promote, this particular type of development. This argument was made in our original 2012 paper (see Table 2), wherein it was demonstrated that, although there were numerous "opportunities" for "training in the responsible conduct of research", *none* of these opportunities was actually aligned with the KSAs of ethical reasoning. This is not surprising since none was *designed* to be aligned with these KSAs; the point is that there are *no KSAs* to which they **are** aligned (see Schmaling and Blume 2009 for discussion of the result of content-only ethics education that does not include reasoning). The only "learning opportunities" identified by the institution (Tractenberg and FitzGerald 2012, Table 2) were passive- and those were generally limited to graduate students. All of the other "training opportunities"—all passive—were also all general, lacking any specificity for either the level of the participant or their development of any particular skill to any particular criterion, much less in any specific discipline. These challenges were identified and discussed in that 2012 paper, and they are included here only to demonstrate how the Mastery Rubric for Ethical Reasoning (2012) can be (and was, in that paper) used to evaluate existing training, and how it *could* be used to create new learning experiences.

We have also discussed how a focus on professionalism and the development of a sense of professional identity, and not on "teaching ethics", is the key to engaging undergraduates as well as graduate students, in the level of ethical reasoning they would need to exhibit as "future professionals" (Tractenberg et al. 2014). In that 2014 paper we explored the professional codes of conduct for statistics and for computing, and integrated our Ethical Reasoning KSAs with an implicit focus on the development of the professional "habits of mind" that are valued by, and exemplified in, experts in those fields (see also Tractenberg 2013; Tractenberg and FitzGerald 2015). These examples are given to support the argument that, even if a single course on ethical reasoning for the discipline is all that can be integrated, its potential to promote professionalism and the development of a professional identity can shift the focus away from "learning ethics" (generic) towards "learning to think like an (ethical) expert or professional in my field" (specific to the discipline). This shift embodies, and promotes, disciplinary stewardship: the engagement of more faculty in this instruction enterprise represents their commitment to stewardship, while the introduction of the key aspects of stewardship—whether for a discipline (Golde 2006) or for (Big) Data (Gates Foundation 2011)—can serve to initiate the students in that degree program to these Stewardly characteristics ("disciplines and fields are dynamic, and *require stewardship*;" "the quality and integrity of disciplines must be actively preserved and conserved;" and, "there are particular habits of mind that characterize 'those to whom we can entrust' the core features of a discipline or field.").

Our ethical reasoning developmental training paradigm is based on the argument that "ethics education should *inculcate*—seed and support the development of—a professional and ethical identity that can then grow over a *career* in science or practice (or both)" (Tractenberg 2016a; see also Tractenberg et al. 2014). Ethical reasoning around Big Data should be perceived—and valued—as representing "professional-level thinking"—rather than as "training required for all students", which tends to diminish its perceived *and its actual* value. 'Ethical reasoning' comprises a learnable, improvable skill set (see Tractenberg et al., in review)—unlike "ethics"—which is an entire field itself, or "responsible conduct of research", which is not a particular skill set, and which is actually defined as varying according to the discipline in some senses (see Steneck 2007: p. xi).

True integration of ethical reasoning into the preparation for participation in/with Big Data will involve an orienting course that presents the opportunities to learn, practice with, and develop facility with, the KSAs of ethical reasoning—and familiarization with the complexities of ethical challenges with Big (and small) Data from within their discipline (see e.g., Tractenberg, this volume). This integration supports training objectives around the preparation for participation to work in/with Big Data, but the focus on *reasoning* rather than specific challenges we face—or are emerging- **now** also promotes general (important) thinking and metacognitive skills (Tractenberg et al., in review). The stewardship model can support engagement by faculty across disciplines, and can also strengthen the motivation for students (and faculty) to engage and fully as possible with a developmental approach to stewardly, professional, ethical reasoning and its use in and with Big Data.

Acknowledgments The Author was supported by a grant (Award 1237590) from the National Science Foundation.

References

American Anthropological Association. http://www.aaanet.org/issues/policy-advocacy/upload/AAA-Ethics-Code-2009.pdf.

American Society of Mammalogists. http://www.mammalsociety.org/uploads/ASM%20Ethics%20Statement.pdf.

Association of Internet Researchers Ethics Committee. (2012). *Ethical decisionmaking and internet research.* http://aoir.org/reports/ethics2.pdf downloaded December, 5 2013.

Biggs, J., & Tang, C. (2007). *Teaching for quality learning at university: What the student does, 3E.* Berkshire, England: McGraw Hill.

Bill and Melinda Gates Foundation. (2011). *Global health data access principles.* Accessed from https://docs.gatesfoundation.org/Documents/data-access-principles.pdf on June, 16 2015.

Bok, D. (1974). On the purposes of undergraduate education. *Daedalus 103*(4), *American Higher Education: Toward an Uncertain Future*, Volume I (Fall 1974), pp. 159–172. Downloaded from http://www.jstor.org/stable/20024257 on July, 19 2015.

Collins, F. S., & Tabak, L. A. (2014). Policy: NIH plans to enhance reproducibility. *Nature, 505*, 612–613, January, 30 2013.

DeMartino, G. F. (2011). *The economist's oath: On the need for and content of professional economic ethics.* New York, NY: Oxford University Press.

Freedman, D. H. (2010). *Lies, damned lies, and medical science.* The Atlantic November 2010 issue; downloaded from http://www.theatlantic.com/magazine/archive/2010/11/lies-damned-lies-and-medical-science/308269/ on November, 7 2015.

Golde, C. M. (2006). Preparing stewards of the discipline. In C. M. Golde., G. E. Walker (Eds.), *Envisioning the future of doctoral education* (pp. 3–20). San Francisco, CA: Jossey-Bass.

Golde, C. M., & Walker, G. E. (Eds.). (2006). *Envisioning the future of doctoral education.* San Francisco, CA: Jossey-Bass.

Hollander, R., & Arenberg, C. R. (Eds.). (2009). *Ethics education and scientific and engineering research.* Washington: National Academy of Engineering.

Kligyte, V., Marcy, R. T., Waples, E. P., Sevier, S. T., Godfrey, E. S., Mumford, M. D., & Hougen, D. F. (2008). Application of a sensemaking approach to ethics training in the physical sciences and engineering. *Science and Engineering Ethics, 14*(2), 251–278.

Lee, L. M., McCarty, F. A., & Zhang, T. R. (2015). Ethical numbers: Training in US graduate statistics programs, 2013–2014. *The American Statistician, 69*(1), 11–16. doi:10.1080/00031305.2014.997891.

McNutt, M. (2014). Editorial: Journals unite for reproducibility. *Science, 346*(6210), p. 679, November, 7 2014. doi: 10.1126/science.aaa1724.

Messick, S. (1994). The interplay of evidence and consequences in the validation of performance assessments. *Educational Researcher, 23*(2), 13–23.

National Institutes of Health (NIH). (2009). *Update on the requirement for instruction in the responsible conduct of research*. November 24. Available online at http://grants.nih.gov/grants/guide/notice-files/NOT-OD-10-019.html.

Santa Clara University. (no date). *Ethical reasoning*. Downloaded from http://www.scu.edu/ethics/ November, 29 2009.

Schmaling, K. B., & Blume, A. W. (2009). Ethics instruction increases graduate students' responsible conduct of research knowledge but not moral reasoning. *Accountability in Research, 16*, 268–283.

Shulman, L. S. (2008). Foreward. In G. E. Walker, C. M. Golde, L. Jones, A. C. Bueschel, & P. Hutchings (Eds.), *The formation of scholars: Rethinking doctoral education for the twenty first century* (pp. ix–xiii). San Francisco, CA: Jossey-Bass.

Steneck, N. H. (2007). *ORI introduction to the responsible conduct of research, revised*. Accessed from https://ori.hhs.gov/sites/default/files/rcrintro.pdf, February 2010.

Tractenberg, R. E. (2013). Ethical reasoning for quantitative scientists: A mastery rubric for developmental trajectories, professional identity, and portfolios that document both. In *Proceedings of the 2013 Joint Statistical Meetings*, Montreal, Quebec, Canada.

Tractenberg, R. E. (2016a). Integrating the ASA's ethical guidelines for professional practice into course, program, and curriculum. In J. Collmann & S. Matei (Eds.), *Ethical reasoning in big data*. New York: Springer.

Tractenberg, R. E. (2016b). Creating a culture of ethics in biomedical big data: Adapting 'guidelines for professional practice' to promote ethical use and research practice. In L, Floridi., B. Mittelstadt (Eds.), *Ethics of Biomedical Big Data*. London: Springer.

Tractenberg, R. E., & FitzGerald, K. T. (2012). A mastery rubric for the design and evaluation of an institutional curriculum in the responsible conduct of research. *Assessment and Evaluation in Higher Education, 37*(7–8), 1003–1021.

Tractenberg, R. E., & FitzGerald, K. T. (2015). Responsibility in the conduct of quantitative sciences: Preparing future practitioners and certifying professionals. (Presented at 2014 Joint Statistical Meetings, Boston, MA). In *Proceedings of the 2015 Joint Statistical Meetings*. Seattle, WA.

Tractenberg, R. E., McCarter, R. J., & Umans, J. (2010). A mastery rubric for clinical research training: guiding curriculum design, admissions, and development of course objectives. *Assessment and Evaluation in Higher Education, 35*(1), 15–32. doi:10.1080/02602930802474169.

Tractenberg, R. E, Russell, A., Morgan, G., FitzGerald, K. T., Collmann, J., Vinsel, L., Steinmann, M., & Dolling, L. M. (2014). Amplifying the reach and resonance of ethical codes of conduct through ethical reasoning: Preparation of Big Data users for professional practice. *Science and engineering ethics*. http://link.springer.com/article/10.1007%2Fs11948-014-9613-1.

Trede, F. (2012). Role of work-integrated learning in developing professionalism and professional identity. *Asia-Pacific Journal of Cooperative Education, 13*(3), 159–167.

Tractenberg, R. E., FitzGerald, K. T., & Collmann, J. (in review). Evidence of sustainable learning with the Mastery Rubric for ethical reasoning.

Walker, G. E., Golde, C. M., Jones, L, Bueschel, A. C., Hutchings, P. (2008). *The formation of scholars: Rethinking doctoral education for the twenty-first century*. San Franciscso, CA: Jossey-Bass.

Wiggins, G., & McTighe, J. (1998). What is backward design? In *Understanding by design*. Upper Saddle River, NJ: Merrill Prentice Hall (pp. 7–19). Retrieved from http://nhlrc.ucla.edu/events/startalkworkshop/readings/backward-design.pdf.

Zoological Association of America. http://www.zaa.org/pdf/ZAA%20Code%20of%20Conduct%20-FINAL-%202.9.09-old.pdf.